D0891447

PREDICTION OF THE ENVIRONMENTAL FATE OF CHEMICALS

PREDICTION OF THE ENVIRONMENTAL FATE OF CHEMICALS

YUSAF SAMIULLAH

Mott MacDonald Environmental Consultants Ltd, Croydon, UK

Formerly, *Monitoring and Assessment Research Centre, King's College London, University of London, UK*

Published in association with
The British Petroleum Company p.l.c.

ELSEVIER APPLIED SCIENCE
LONDON and NEW YORK

ELSEVIER SCIENCE PUBLISHERS LTD
Crown House, Linton Road, Barking, Essex IG11 8JU, England

Sole distributor in the USA and Canada
ELSEVIER SCIENCE PUBLISHING CO., INC.
655 Avenue of the Americas, New York, NY 10010, USA

WITH 55 TABLES AND 54 ILLUSTRATIONS

British Library Cataloguing in Publication Data

Samiullah, Yusaf
Prediction of the environmental fate of chemicals.
1. Environment. Pollution by chemicals. Pollution of environment by chemicals. Simulations. Models
I. Title II. British petroleum company
628.5

ISBN 1-85166-450-5

Library of Congress Cataloging-in-Publication Data

Samiullah, Yusaf.
Prediction of the environmental fate of chemicals/Yusaf Samiullah.
p. cm.
'Published in association with the British Petroleum Company.'
Includes bibliographical references.
ISBN 1-85166-450-5
1. Pollution—Environmental aspects. I. British Petroleum Company. II. Title.
TD196.C45S36 1990
628.5′2—dc20 89–48129
CIP

Printed in Northern Ireland by The Universities Press (Belfast) Ltd.

Foreword

Concern over the effects of chemicals in the environment has been increasing for many years. Environmental contamination by DDT, Aldrin, Dieldrin, mercury, PCBs, organotins and many other substances are all part of the public consciousness and have led to widespread attention to this topic. Some of the concerns have arisen because human health has been affected when contaminants have been consumed via the food chain—for instance in the case of 'Minimata disease' in Japan. In other cases, direct effects on other components of ecosystems have given cause for alarm.

The toxic effects which any chemical can cause are a function of exposure and innate toxicity, i.e. of the ability to reach in sufficient quantity a site where a biological process can be disrupted and of the tendency to cause disruption when it gets there.

The processes by which chemicals reach sites of toxic action are the subject of this book, and are a fundamental consideration in ecotoxicology. When a chemical enters the environment e.g. via a spillage or in an effluent, it is potentially subject to a wide variety of processes which may eliminate it from the environment completely, modify it into a more or less harmful substance, or transfer it to another part of the environment. The processes involved are complex and highly variable, but it is essential to increase our understanding of them.

If the toxic effects of chemicals are to be controlled it is essential that the way that these processes occur can be predicted. Only then can the probable environmental hazards posed by new chemicals be assessed and appropriate control measures defined before the substance reaches the environment.

The need for methods to predict the environmental hazards posed by chemicals is internationally recognized, and various regulatory schemes are in place to ensure that relevant data is gathered before the chemical is marketed. An essential element of such schemes is the ability to predict fate and behaviour of chemicals, often based on limited experimental data.

This book sets out some of the methods and principles which can be used in predicting and understanding the fate of chemicals in the environment. It is based on a report initially produced by MARC for BP, which was intended to summarize and review the data available. We are pleased to have been able to update the information and to publish it as a contribution to the development of a complex, but important, topic.

DR D. C. MONK
BP International Ltd

Acknowledgement

The author is grateful to Group Environmental Services, The British Petroleum Company, London, for sponsorship towards the preparation of this book under contract numbers XEC-935-176-1 and XEC-935-287-1/ZL47.

Contents

Chapter 1

Introduction

This book was commissioned to review the state of the art of predicting the environmental fate of chemicals. It is intended to be a significant bibliographic resource, highlighting major developments in empirical and theoretical approaches to fate prediction. A very wide range of topics are considered, the intention being to provide the reader with insight into the nature of the task that may be required. Coverage of any particular subject may be expanded by reference to the appropriate paper(s) cited in the References and Bibliography.

Other than a means of recording present-day achievements and ambitions for chemical fate prediction in the air, water and soil, the book should provide sufficient information to enable science students, scientists, municipal administrators, politicians and the environmentally concerned general public to ask pertinent questions of those purporting to be experts in chemical fate prediction. Thus, given a list of potentially useful physicochemical parameters it should be possible to request not only that a commissioned study should incorporate some such parameters, but also that the study should justify the exclusion of others.

A primary aim of the book is to outline what it is about a chemical that determines where it will go in the environment if released. Inevitably it is impossible to consider a chemical in isolation—the medium to which it may be released and with which it may interact is an integral part of the fate prediction equation. Emphasis is placed on the final environmental medium to which a substance will migrate. The enormous area of chemical toxicity to living organisms is treated only in so far as bioconcentration can be anticipated.

Ideally, environmental fate prediction requires the acquisition of three sets of data. Firstly, information is required on the physical and chemical properties of the substance under consideration. Secondly, data need to be collected on the routes of transfer to and through the environment for that substance. Finally, the resulting distribution of the substance in the various environmental phases should be ascertained. In practice, however, our current knowledge may cover, and then only partially, one or two of these sets of data. Predicting the environmental fate of chemicals may therefore call for extrapolation from confirmed fact to speculative distribution. Conversely, a knowledge of environmental distribution can itself provide clues as to the physical and chemical properties of molecules and yield information on likely environmental transfer and transformation processes.

The prediction of the environmental fate of chemicals may therefore be considered as an integrated process utilizing both preproduction information on chemical structure and properties, and postproduction information about the distribution of that chemical in the environment. An information hierarchy can be envisaged, with data on fundamental molecular properties of structure and energy being used to predict molar or bulk chemical properties dependent on molecular aggregations. Some molar properties, such as solubility, vapour pressure and partition coefficient, may be used, in conjunction with knowledge of natural chemical and physical pathways in the environment, to anticipate eventual environmental fate. For instance, the rate of volatilization of a chemical may be deduced from Henry's Law constant which in turn is derived from information on solubility and vapour pressure. These last two parameters are themselves calculable from a knowledge of molecular structure and boiling temperature. The relationships between molecular and molar properties of chemicals and the prediction of volatilization rate are illustrated in Fig. 1.1. Information on potential marketing trends, possible use, transport, storage and disposal of chemicals may provide a more obvious means of establishing primary sites of environmental contamination.

Further down the hierarchy, the environment itself may be conceptually split into three components, viz. the soil, aquatic and atmospheric phases. Prediction of ultimate chemical fate may require the subdivision of these units into smaller, and mathematically more manageable, portions. Such a multicompartmental construction can then be used to indicate data requirements for modelling studies. Data may simply be chemical concentrations measured in a number of media, or may involve quantified rates of chemical transfer between compartments or suppositions about degradative processes.

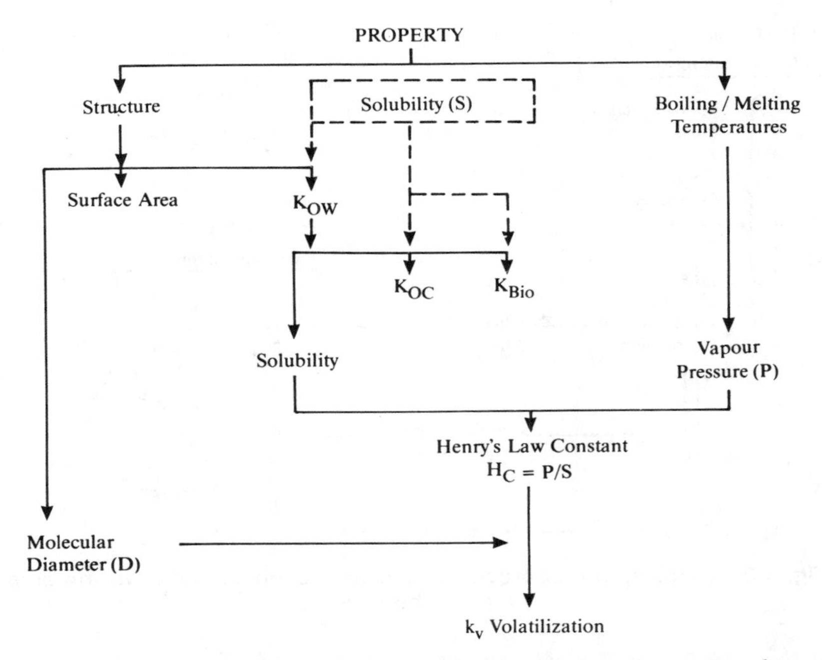

Fig. 1.1. The relationship between molecular and molar properties of chemicals and physical transport constants. K_{OW}, Water/octanol partition coefficient; K_{OC}, water/soil or sediment partition constant; K_{Bio}, water/biota partition coefficient. Reproduced from Mill (1981).

Once equilibrium has developed, chemical distribution in the environment provides a source of information on the properties and behaviour of the chemical. Thus, a two-way information transfer mechanism exists within the conceptual hierarchy. Figure 1.2 outlines these concepts and illustrates the hierarchical nature of environmental fate prediction. This figure also provides a guide to the rationale behind the structure of the book. A decision as to whether data need be collected on all or only some of the stages indicated should be determined by present knowledge of chemical behaviour and properties, environmental distribution and persistence, and potential hazard to biological systems.

This book begins with a chapter on the fundamental properties of chemicals. Molecular connectivity is commended both as a way of utilizing data on fundamental molecular properties and predicting unknown molecular and molar properties. Deemed the most useful of

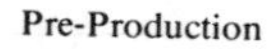

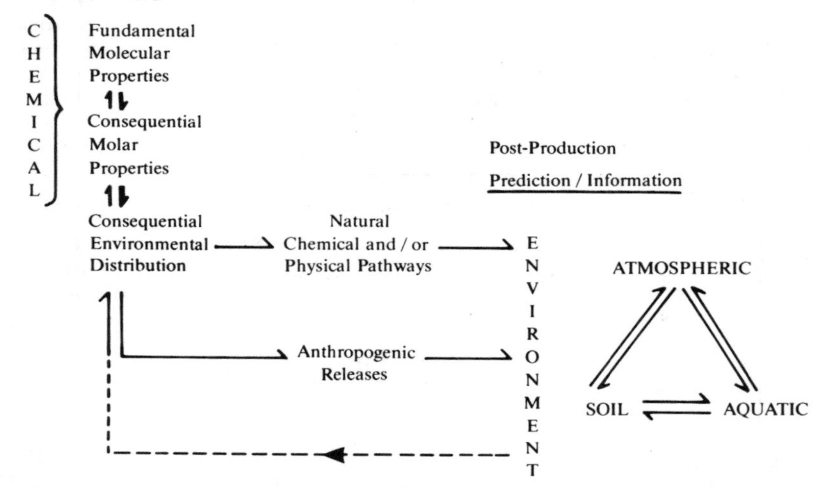

Fig. 1.2. A conceptual approach to the prediction of the environmental fate of chemicals.

the consequential molar properties, partition coefficients are discussed in detail. Natural chemical and physical pathways and transfers are outlined, as are potential anthropogenic releases of synthetic chemicals, be these associated with manufacturing processes or hazardous waste disposal. Next, various modelling strategies and philosophies are covered and, by way of recommendation, a whole section is devoted to modelling procedures utilizing the concept of fugacity. The natural world is discussed in three chapters on soil, aquatic and atmospheric environments. Intercompartmental transfers and significant degradative processes are detailed within the relevant chapter.

Within the context of specific equations, the book indicates data requirements throughout the various chapters. Wherever appropriate, published theory is cited and examples of existing fate predictive models are presented. Notwithstanding the apparent complexity of environmental fate prediction, it is hoped that it has been demonstrated that great progress in our understanding of the processes involved has been made over the last two decades, and that whilst data may be as yet unavailable, at least we are now in a position to recognize that fact and take steps to rectify omissions.

Chapter 2

Fundamental Properties of Chemicals

2.1 INTRODUCTION

There are over two million known chemicals of which some 70 000 are presently in commercial use worldwide. These include metals, metal compounds, metalloids and an enormous variety of synthetic organic chemicals. Testing a new chemical for its likely adverse biological effects may cost from a few thousand US dollars for simple acute toxicity tests or laboratory measurement of bioconcentration potential to over one million US dollars for longer-term epidemiological studies on humans. Even then, in only 20% of all cases can health hazard assessments fulfil rigorous scientific and technical criteria (NRC, 1984). Furthermore, such testing procedures may not account adequately for the complex series of transformations undergone by chemicals once they are subjected to environmental processes. Thus, it is clear that some form of early screening is necessary, even if only to identify those chemicals likely to require further testing.

A logical point at which to begin such a first check is to consider those aspects of chemical property and structure amenable to theoretical prediction. These may be considered in two distinct yet interrelated groups, namely molecular properties and molar properties. Molecular properties are independent of state and are intrinsic to the single molecule and include such fundamental parameters as the heat of formation (ΔH_f) and heat of atomization (ΔH_a). Molar properties, however, depend on molecular aggregations, examples being the heat of vaporization (ΔH_{vap}) boiling point, liquid density or water solubility.

The behaviour of chemicals in soil, water or the atmosphere is ultimately determined by the interaction of the molecular or molar forces and properties of the substance with those of the relevant medium. Therefore, what has to be the goal in the theoretical fate prediction is a measure of the probability that a given property will behave in a certain way towards a defined abiotic or biotic receptor.

Since chemical properties are a consequence of molecular structure and electronic charge, many studies have been carried out into the relationship between chemical activity and structure. This field of endeavour has now developed into what are generally termed investigations into Quantitative Structure–Activity Relationships (QSAR) and these are widely used in chemistry in the prediction of both properties and the course of reactions.

The first part of this chapter provides a brief introduction to the background to studies demonstrating the utility of structure–activity relationships (SAR), representative of individual molecules or molar aggregations. Their role as a predictive tool in toxicology and ecotoxicology has been reviewed comprehensively by Goldberg (1983), ECETOC (1986) and Nirmalakhandan and Speece (1988). In spite of the apparent potential of QSAR predictions, it should always be borne in mind that these are only as good as the biological and chemical data upon which they are based. In addition, particularly when considering predictions of toxicity, the effects of metabolic intermediates, animal species differences, synergistic and antagonistic mixtures of substances, and individuals susceptibilities, are often not accounted for in the initial data set.

The chapter then concentrates on the partition coefficient (K) which has proved to be one of the most useful molar parameters both in environmental fate prediction and in the estimation of bioconcentration potential. Ways of determining K are discussed, and examples are given of the use of partition coefficient in bioconcentration factor studies.

2.2 MOLECULAR AND MOLAR PROPERTIES

2.2.1 Molecular structure

The concept that molecules consist of atoms bound together into stable and identifiable entities has played a vital role in modern chemistry. Physical properties, stability, reactivity and other characteristics are described and explained in molecular terms.

At the moment, quantum mechanics offers the most fundamental approach to the quantification of molecular structure. In principle, given the coordinates and atomic numbers for a collection of atoms, the Schroedinger equation can be solved for the eigen values and eigen vectors that describe energy and electron distribution. The stable arrangement of these atoms in molecular form corresponds to their lowest energy arrangement. Ordinarily, a chemist cannot accurately describe the electronic structure or energies of two isomers (e.g. butane and isobutane) without a quantum mechanical calculation. At a less fundamental level it is possible to deduce the structural formulae of isomers based on notions of valence and chemical bonding. This intermediate level of structural information is the bonding or branching pattern in the molecule and may be classified topologically.

Figure 2.1 shows a hierarchy of structural descriptions for a molecule. At the most primitive level A, information is limited to the type of atoms present in the molecule. Level B conveys information that the molecule is an alkane and allows for other generalizations about chemistry and properties. Level C allows for approximate values of solubility, boiling point and density, as well as chemical reactivity, although a specific molecule cannot be identified due to isomeric possibilities. Level D informs on how the atoms in the molecule are organized or connected. At level E, the quantum mechanical description, all of the information about the molecule is contained in principle.

However, in practice, especially for interacting systems, quantum mechanical calculations have not produced adequate methods for predicting the properties of large molecules. This is due to two main problems. Firstly, because of the complexity of dealing with all the interactions between particles as required by the Schroedinger equation, there are practical problems of computation and computer time. Secondly, a quantity such as total bonding energy is only a tiny fraction of the total calculated energy of a molecular system. A small but acceptable error in terms of the total energy becomes a major portion of the chemical bonding energy. For example, even a stable diatomic molecule with a strong bond such as N_2 (945 kJ (225 kcal)) has a total molecular energy of more than 420 000 kJ (100 000 kcal) (Kier & Hall, 1976). It is not feasible to ask for a calculation based on the Schroedinger equation to yield results better than to within one part per 100 000. Thus, for such reasons, more efforts have been directed at less complex descriptions than those of quantum mechanics.

A. C, H

B. C_nH_{2n+2}

C. C_4H_{10}

D. $CH_3-CH_2-CH_2-CH_3$ →

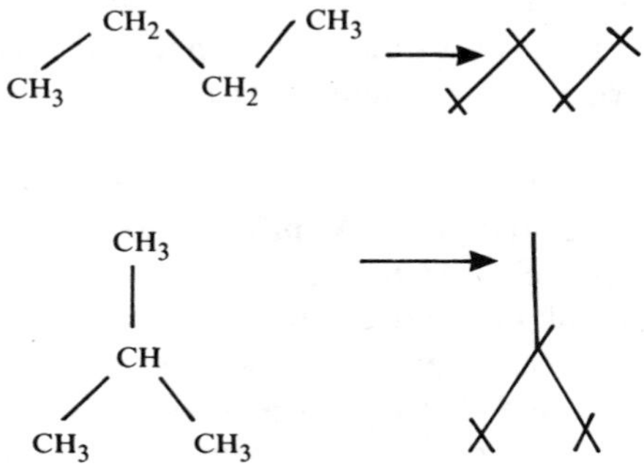

Numerical Indices

$CH_3-CH(CH_3)-CH_3$ →

E. $\psi_1 = c_1\phi_1 + c_2\phi_2 + \ldots\ldots$

$\vdots$

$\psi_n = c'_1\phi'_2 + c'_2\phi'_2 \ldots\ldots$

Fig. 2.1. A hierarchy of structural descriptions.

2.2.2 Additive and constitutive properties

Molecular properties may be divided into two general classes. Those such as molecular mass, which may for all practical purposes be obtained as a sum of the corresponding values for the constituent parts, are termed additive, whereas properties that depend heavily on details of the arrangement of the constituent atoms are termed constitutive. Boiling point, for instance, is constitutive, as is the water solubility of organic compounds.

Molecular weight is strictly additive in terms of the numbers and kinds of atoms in the molecule, as is demonstrated in Table 2.1. This additivity is fundamental to the concept of homologous series in organic chemistry.

Table 2.1
Structural influences on selected properties of alkane homologous series

Compound	*Heat of atomization*[a]	*Molar refraction*[b]	*Molar volume*[c]	*Molecular weight*	*Refractive index*[d]	*Boiling point*[d]	*Specific gravity*[d]
Butane	1 234·96	—	—	58·13	—	−5·0	—
Pentane	1 514·80	25·27	115·22	72·15	1·357 5	36·07	0·626 2
Hexane	1 794·72	29·91	130·68	86·17	1·374 9	68·74	0·659 4
Heptane	2 074·75	34·54	146·52	100·19	1·387 6	98·43	0·683 8
Octane	2 354·86	39·19	162·58	114·21	1·397 4	125·67	0·702 5
Nonane	2 634·76	43·83	178·69	128·23	1·405 4	150·81	0·717 6
Decane	2 914·84	48·47	194·84	142·25	1·411 9	174·12	0·730 1

[a] Heat of atomization in kcal mol^{-1} taken from Cox and Pilcher (1970) (Table 34).
[b] Molar refraction R_m calculated as $[(n^2-1)/(n^2+1)](M/d)$, in $cm^3\ mol^{-1}$, when n is refractive index and d is density.
[c] Molar volume V_m calculated as M/d, where M is molecular weight and d is density.
[d] Data taken from *Handbook of Tables for Organic Compound Identification*, CRC Press, Cleveland, OH.
Table reproduced from Kier and Hall (1976).

The heat of atomization in the hydrocarbon series is additive within experimental error, correlating nearly with the number of carbon atoms. Molar volume and molar refraction can also be seen to be additive in the series chosen for Table 2.1. However, for the properties of boiling point and specific gravity, the increment between successive members of the homologous series slowly decreases, demonstrating a nonlinear correlation with the number of carbon atoms.

Kier and Hall (1976) pointed out that the relationship between molecular structure and properties is less clear when branched molecules are considered. For example, with the isomeric hexanes, none of the properties cited in Table 2.1 are the same for any two isomers. Clearly, knowledge of the number of atoms in the molecule is insufficient to describe all the salient features of structure that govern the magnitude of a property. What is required is sufficient information about structure in order that correlation may be made with physicochemical properties. Molecular connectivity, which is concerned with the quantification of the topology of molecules, endeavours to convey such structural information.

2.2.3 The quantification of molecular structure

In order to quantify molecular structure at the topological level, numerical descriptors were developed, encoding within them information relating to the number of atoms and their arrangement, in other words, their connectivity. Such structural descriptors should correlate with properties dependent on molecular connectivity. Kier and Hall (1976) set out the following criteria as guidelines in the evaluation of a method for the quantification of molecular connectivity:

A. The method should serve as a basis of structural definition with the capacity for wide application to many physical, chemical, and where possible, biological properties.

B. The approach should be built upon fundamental principles of molecular structure rather than upon empirical quantities.

C. The method should make use of simple computations that are not time consuming, but none the less readily computerized.

D. The numerical descriptors should be unique for a given structure.

E. For practical application, the method must be sufficiently flexible to handle such structural features as heteroatoms, unsaturation, cyclization and aromaticity. Applications should not involve the extensive addition of new parameters or features greatly increasing the number of operations or tedium in calculation.

F. The method should lend itself to amalgamation with certain indices derived from quantum mechanical approaches. Such a combination of topological and molecular orbital indices may provide a powerful tool in the development of structure–activity relationships.

In the context of molecular connectivity, a diagram of molecular structure is termed a graph. In the topological approach, emphasis is given to the *fact* rather than the type of connection.

The graph for methane is constructed by using points for vertices (atoms) and lines connecting points for edges (bonds).

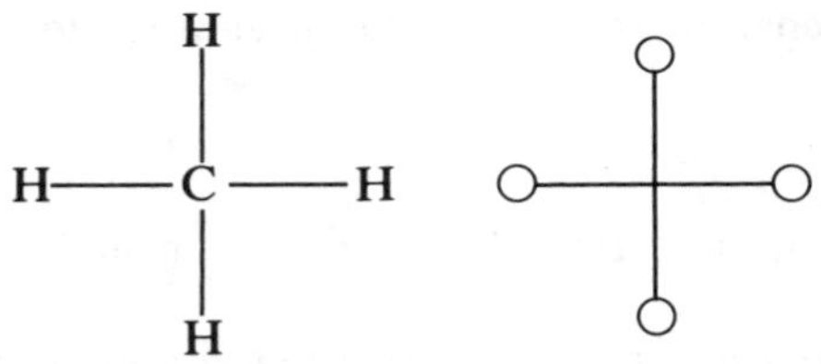

The graph for ethylene requires two edges connecting a pair of vertices. These are called multiple edges and are drawn as curved or straight lines.

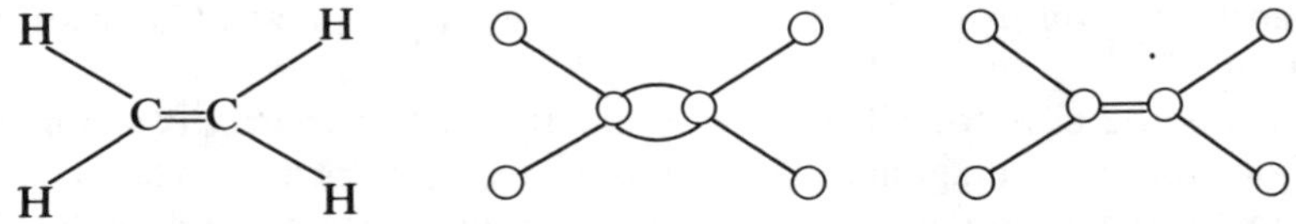

No edge begins and ends at the same vertex. A graph symbolized by G may be composed of vertices V(G) and edges E(G), Thus, the total graph should be designated G[V(G), E(G)], where E(G) is a finite set of edges and V(G) a finite set of vertices. The symbol n designates the number of vertices and the symbol m the number of edges.

Sylvester (1874) is credited with first showing that a chemical graph could be written as a matrix with entries for edges contributing the

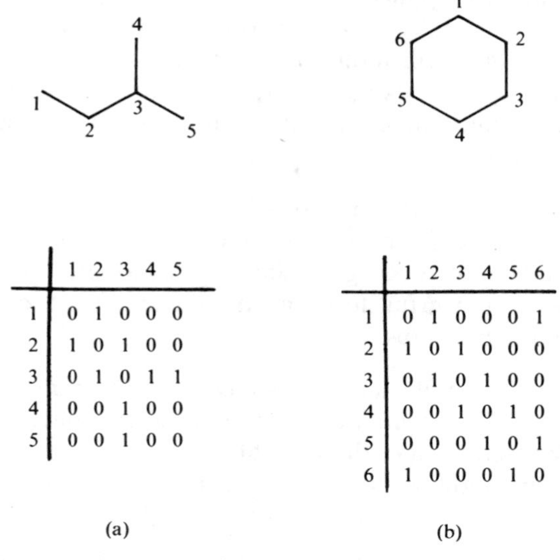

	1	2	3	4	5
1	0	1	0	0	0
2	1	0	1	0	0
3	0	1	0	1	1
4	0	0	1	0	0
5	0	0	1	0	0

(a)

	1	2	3	4	5	6
1	0	1	0	0	0	1
2	1	0	1	0	0	0
3	0	1	0	1	0	0
4	0	0	1	0	1	0
5	0	0	0	1	0	1
6	1	0	0	0	1	0

(b)

Fig. 2.2. The transformation from chemical graph to topological matrix.

nonzero terms in the array. This array is called the *topological* or *adjacency* matrix.

In order to construct this matrix, the graph is numbered in any order. Vertex numbers correspond to the row and column designations of the matrix. An entry T_{ij} in the matrix has the value one when there is an edge between vertices $_i$ and $_j$; otherwise it is zero. The transformation from chemical graph to topological matrix is illustrated in Fig. 2.2.

As is the case for all topological matrices, this matrix is symmetric about the main diagonal. The valence or degree of each vertex is the sum of the unit entries in a row or column corresponding to that particular vertex. Thus, in Fig. 2.2(a) the second vertex from the left has two entries in the second row (or column) in the matrix. This is the valence of the vertex. In Fig. 2.2(b) all vertices have a valence of 2, corresponding to two unit entries in all rows and columns of its topological matrix. The formulation of algebraic expressions based on the topological matrix has been discussed further by Kier and Hall (1976).

2.2.4 The development of an index of molecular connectivity

The reduction of a graph or topological matrix to a topological index, which more readily offered quantitative comparison with physical or chemical properties of molecules had long been desired. Cited in A to F are significant contributions made towards this goal:

A. The Wiener Path Number (Wiener, 1947, 1948*a*).
B. Altenburg Polynomial (Altenburg, 1961, 1966).
C. The Gordon and Scantlebury Index (Gordon & Scantlebury, 1964).
D. The *Z* Index of Hosoya (Hosoya, 1971).
E. The Smolenski Additivity Function (Smolenski, 1964).
F. The Randic Branching Index (Randic, 1975).

Randic (1975) proposed a scheme built around a unique numbering of atoms in a hydrocarbon series, with the selection of a particular numbering scheme based on an analysis of the corresponding adjacency matrix. Figure 2.3 uses propane as an example of the several ways of numbering the vertices. If the rows of the matrix are considered to be binary numbers, three values can be written for each matrix. Interpreting these in decimal notation results in three different numbers. Randic selected the numbering scheme of the hydrocarbon with the lowest decimal number and used that number as an index associated with that particular hydrocarbon.

Emerging from the quest for a suitable topological index, the molecular connectivity index χ became a descriptor of molecular structure. It is based on a count of groupings of skeletal atoms and the number of valence electrons of each skeletal atom, and is weighted by the degree of skeletal branching. For a formal exposition of the method of calculating molecular connectivity indices, reference should be made to Kier and Hall (1976). In brief, however, the following procedure outlined by Koch (1982*b*) should suffice to explain the stages involved.

Calculation of the first-order and the second-order valence molecular connectivity index [($^1\chi^v$) and ($^2\chi^v$)]

Generally, in the connectivity method, organic molecules are represented by a molecular skeleton and each moiety stands for the skeletal atom together with any bonded hydrogen atoms. Each of these groups is represented by a delta value (δ). The valence delta value (δ^v) represents the difference between the valence electrons

Numbering Scheme	Adjacency Matrix	Binary Equivalent	Decimal Equivalent
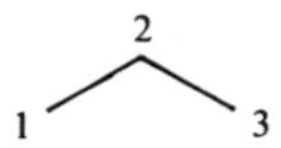 1, 2, 3	$\begin{pmatrix} 0 & 1 & 0 \\ 1 & 0 & 1 \\ 0 & 1 & 0 \end{pmatrix}$	10, 101, 10	252
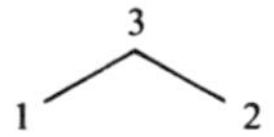 1, 3, 2	$\begin{pmatrix} 0 & 0 & 1 \\ 0 & 0 & 1 \\ 1 & 1 & 0 \end{pmatrix}$	1, 1, 110	116
 3, 2	$\begin{pmatrix} 0 & 1 & 1 \\ 1 & 0 & 0 \\ 1 & 0 & 0 \end{pmatrix}$	11, 100, 100	344

Fig. 2.3. Numbering scheme for the hydrogen-suppressed graph of propane along with the corresponding adjacency (topological) matrices and their interpretations as binary and decimal numbers. Reproduced from Kier and Hall (1976).

(Z^v) minus the number of bonded hydrogens (h) (eqn (1)):

$$\delta^v = Z^v - h \tag{1}$$

e.g. Z^v for C = 4, N = 5, O = 6

For atoms in higher rows of the periodic table, the valence delta value also takes into account the nonvalence electrons (z) and will be calculated in eqn (2):

$$\delta^v = \frac{Z^v - h}{z - Z^v} \tag{2}$$

The valence delta values for three halogens are as follows:

$$\text{Cl} = 0{\cdot}7, \qquad \text{Br} = 0{\cdot}25 \quad \text{and} \quad \text{I} = 0{\cdot}15$$

On this basis a delta value based on eqn (1) is calculated for each atom of the molecule (Fig. 2.4(a)). For the calculation of the ($^1\chi^v$)

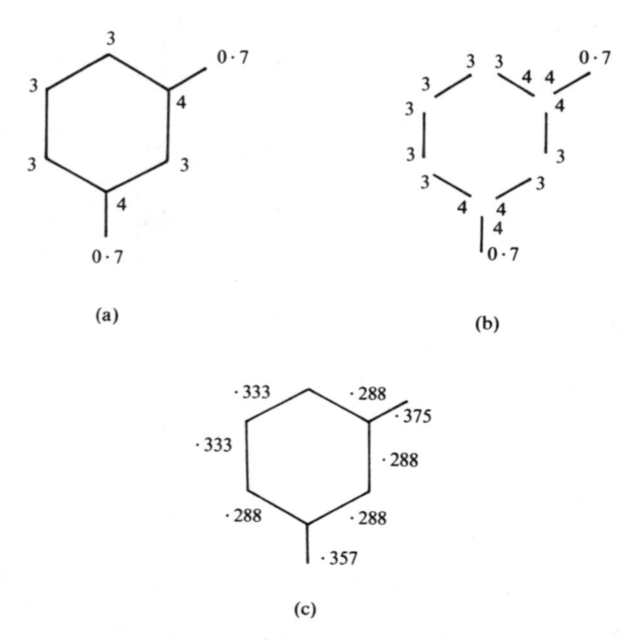

Fig. 2.4. Scheme for calculating the first-order valence molecular connectivity index for dichlorobenzene. Reproduced from Koch (1982*b*).

index, the skeletal structure must be dissected into first-order (one-bond) fragments, each defined by two delta valence values (Fig. 2.4(b)).

For each fragment one can calculate a fragment index according to the algorithm (eqn (3)) (Fig. 2.4(c)):

$$C_{ij} = (\delta_i^v \delta_j^v)^{-0.5} \tag{3}$$

and sum the fragment index values to obtain the first-order valence molecular connectivity index (eqn (4)):

$$^1\chi^v = \sum (\delta_i^v \delta_j^v)^{-0.5} \tag{4}$$

$$^1\chi^v = 2(3 \times 3)^{-0.5} + 4(3 \times 4)^{-0.5} + 2(4 \times 0.7)^{-0.5} = 2.532$$

The second-order valence molecular connectivity index can be calculated on the basis of eqn (5):

$$^2\chi^v = \sum (\delta_i^v \delta_j^v \delta_k^v)^{-0.5} \tag{5}$$

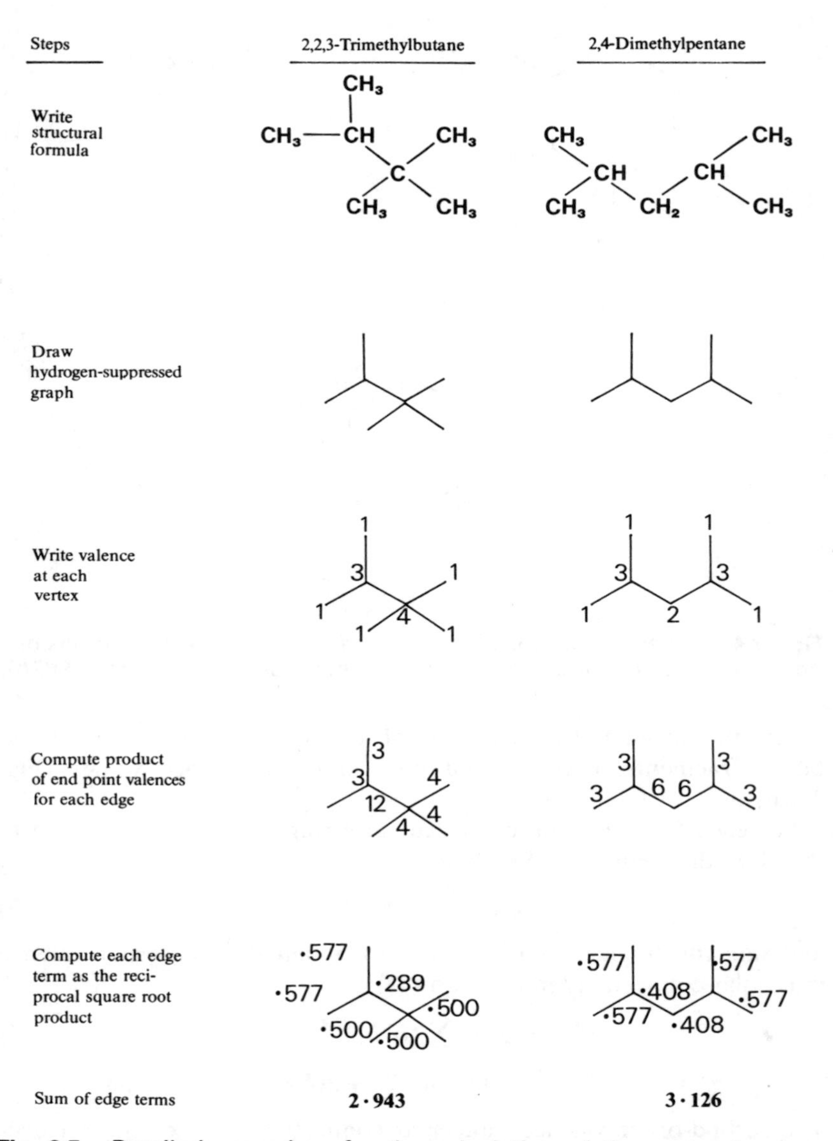

Fig. 2.5. Detailed procedure for the calculation of the molecular connectivity index. Reproduced from Kier and Hall (1976).

Figure 2.5 illustrates the major steps in the calculation of the index for 2,2,3-trimethylbutane and 2,4-dimethylpentane.

2.2.5 Applications of the molecular connectivity index

The predictive potential of the molecular connectivity index is amply demonstrated by correlation with observed molecular and molar properties of chemicals. For example, the heat of atomization for hydrocarbons (ΔH_a) is calculable with the connectivity function. Table 2.2 illustrates the point for part of an alkane series. The different statistical correlations achieved with more complex terms of the connectivity function are presented in Table 2.3.

As can be seen from Tables 2.2 and 2.3, the heat of atomization bears a significant relationship to the number of first-row atoms n.

Table 2.2
Observed and calculated heat of atomization ΔH_a for alkanes from correlation with the connectivity function

Compound	ΔH_a	
	Obs.	*Calc.*
1. Propane	955·49	956·01
2. *n*-Butane	1 236·31	1 236·12
3. 2-Methylpropane	1 238·31	1 238·32
4. *n*-Pentane	1 516·65	1 516·55
5. 2-Methylbutane	1 518·57	1 518·32
6. 2,2-Dimethylpropane	1 521·32	1 521·23
7. *n*-Hexane	1 797·10	1 797·17
8. 2-Methylpentane	1 798·80	1 798·71
9. 3-Methylpentane	1 798·16	1 798·23
10. 2,2-Dimethylbutane	1 801·49	1 801·34
11. 2,3-Dimethylbutane	1 799·63	1 799·95
12. *n*-Heptane	2 077·52	2 077·59
13. 2-Methylhexane	2 079·23	2 079·22
14. 3-Methylhexane	2 078·60	2 078·60
15. 3-Ethylpentane	2 077·97	2 077·99
16. 2,2-Dimethylpentane	2 081·91	2 081·19
17. 2,3-Dimethylpentane	2 080·26	2 079·53
18. 2,4-Dimethylpentane	2 080·92	2 080·39
19. 3,3-Dimethylpentane	2 080·81	2 080·82
20. 2,2,3-Trimethylbutane	2 081·59	2 081·96
21. *n*-Octane	2 357·94	2 358·05
22. 2-Methylheptane	2 359·62	2 359·62

Data from Kier and Hall (1976).

Table 2.3
Correlation of alkane heat of atomization ΔH_a with the connectivity function

Statistical data			*Regression coefficients*							*Constant*
r	s	n	$^1\chi$	$^1\chi_P$	$^5\chi_P$	$^5\chi_C$	$^4\chi_{PC}$	$^5\chi_{PC}$	$^6\chi_{PC}$	
0·999 9	1·455	280·53								115·72
>0·999 9	0·960	283·33	−6·321							116·19
>0·999 9	0·464	286·38	−12·46	1·515	—	−2·474	1·142 0	−2·026 0	—	114·38
>0·999 9	0·371	286·15	−12·08	0·920 9	1·504 6	−2·443	0·859 5	−0·499 6	−1·421	114·65

Data from Kier and Hall (1976).

Kier and Hall (1976) suggest that n is a necessary regression variable for the heat of atomization because ΔH_a is linearly dependent on n as well as on the basic experimental quantity ΔH_f^{gas} for hydrocarbons. When $^1\chi$ is added as a regression variable, the standard error decreases by 35% to 0·96. An eight-subgraph term expression yields a standard error of 0·37 kcal, very close to the 0·32 kcal estimated as the experimental error by Somayajulu and Zwolinski (1972). The seven- or eight-term connectivity function produces nearly the same quality fit as the seven-variable function of Somayajulu and Zwolinski (1966), which explicitly takes account of bond terms, bond interaction and nonbonded steric terms.

The connectivity function also correlates well ($r > 0{\cdot}9$) with the heat of formation (ΔH_f) of alkanes, aliphatic alcohols, ethers, thiols and monoolefinic alkenes.

Molar refractivity (MR), which is directly related to the electronic configuration of a molecule, has been positively correlated with molecular connectivity for a wide range of class of organic compounds. Molar refractivity itself is correlated to many physicochemical properties, some of which may be derived theoretically. Nirmalakhandan and Speece (1988) outlined the example of the relationship of electronic polarization in a molecule (ϕ) where:

$$\phi = (3\pi/4)MR$$

MR was calculable using the equation:

$$MR = (n^2 - 1/n^2 + 2)MW/d$$

where MW is molecular weight, d is density and n is refractive index. Being a molecular property, MR is independent of the physical state, which is not the case for n and d which require experimental measurement.

Some gas equation empirical constants may be described in terms of molecular connectivity. These constants take account of deviations from ideal behaviour as expected from

$$P = RT/V$$

where P is the pressure (atmospheres), T is temperature (K), V is molar volume (litres), and R is the molar gas constant (8·3 J mol^{-1} K^{-1}).

Van der Waals' radii and intermolecular forces are examples of such constants. The diamagnetic susceptibilities of alkanes and aliphatic alcohols are also correlatable with connectivity function. Diamagnetic substances, typically organic compounds, are repelled by magnetic fields in contrast to paramagnetic substances which are attracted.

Molar (bulk) properties such as heat of vaporization, boiling point, liquid density, water solubility and partition coefficient depend on molecular topology both directly and indirectly. The form of the appropriate connectivity function varies depending on the molar property under consideration, but close correlation can usually be achieved with a reasonable number of variables. Table 2.4 illustrates observed and predicted values for partition coefficients of various hydrocarbons, which have been correlated with connectivity.

2.2.6 Structure–activity relationships and toxicology

From the late 1970s, enhanced computing techniques began to be utilized to operate in mathematical domains in more than three dimensions. Pattern recognition and cluster analysis are typical of the new approach. To take the example of SAR studies of chemicals with genetic toxicity, it is assumed that molecular structure and biological activity (genetic toxicity) are related. It is further assumed that this relationship can be discovered by applying statistical and pattern recognition methods to a set of tested components, with the inference that the relationship holds true for untested components (Jurs, 1983; Stouch & Jurs, 1985). The computer software Automatic Data Analysis using Pattern Recognition Techniques (ADAPT) system consists of some 60 FORTRAN routines which can be applied to problems concerning several hundred companies at a time (Stuper *et al.*, 1977, 1979). In the SIMCA pattern recognition method developed by Wold (1976), Wold and Sjostrom (1977) and Dunn and Wold (1980), each class of pattern recognition problems was separately modelled with a principal components analysis. Unknown compounds to be classified were fitted to class models and classification was based on the goodness of statistical fit.

With cluster analysis, similar members of a data matrix are grouped hierarchically in order to produce a tree or box cluster. At any level of clustering, the points within a cluster will be similar to each other in terms of the variables used for clustering and will be different from all other members of the data set (Nirmalakhandan & Speece, 1988).

An illustration of cluster analysis is given by Fig. 2.6 in which 53 petrochemicals are classified into toxic and nontoxic groups, using the second-order molecular connectivity index.

Table 2.4
Observed and predicted values for partition coefficient of hydrocarbons correlated with connectivity

	Compound	*log K*	
		Obs.	*Calc.*
1.	*n*-Pentane	2·50	2·54
2.	2-Methylbutane	2·30	2·41
3.	2-Methylpentane	2·80	2·85
4.	3-Methylpentane	2·80	2·88
5.	*n*-Hexane	3·00	2·98
6.	*n*-Heptane	3·50	3·42
7.	2,4-Dimethylpentane	3·10	3·17
8.	*n*-Octane	4·00	3·89
9.	Cyclopentane	2·05	2·17
10.	Cyclohexane	2·46	2·62
11.	Methylcyclopentane	2·35	2·52
12.	Cycloheptane	2·87	3·06
13.	Methylcyclohexane	2·76	2·96
14.	Cyclooctane	3·28	3·50
15.	1,2-Dimethylcyclohexane	3·06	3·33
16.	1-Pentyne	1·98	2·04
17.	1-Hexyne	2·48	2·48
18.	1-Heptyne	2·98	2·93
19.	1-Octyne	3·48	3·37
20.	1-Nonyne	3·98	3·81
21.	1,8-Nonadiyne	3·46	3·31
22.	1,6-Heptadiyne	2·46	2·42
23.	1-Pentene	2·20	2·20
24.	2-Pentene	2·20	2·20
25.	1-Hexene	2·70	2·64
26.	2-Heptene	3·20	3·08
27.	1-Octene	3·70	3·52
28.	4-Methyl-1-pentene	2·50	2·51
29.	1,6-Heptadiene	2·90	2·73
30.	1,5-Hexadiene	2·40	2·29
31.	1,4-Pentadiene	1·90	1·85
32.	Cyclopentene	1·75	1·86
33.	Cyclohexene	2·16	2·31
34.	Cycloheptene	2·57	2·75
35.	Toluene	2·73	2·54
36.	Ethylbenzene	3·15	2·98
37.	Isopropylbenzene	3·66	3·23
38.	*n*-Propylbenzene	3·68	3·42
39.	Diphenylmethane	4·14	4·41
40.	1,2-Diphenylethane	4·79	4·85
41.	Biphenyl	4·04	4·01
42.	*p*-Xylene	3·25	2·90
43.	Benzene	2·13	2·17
44.	Naphthalene	3·37	3·42
45.	Phenanthrene	4·81	4·66

Data from Kier and Hall (1976).

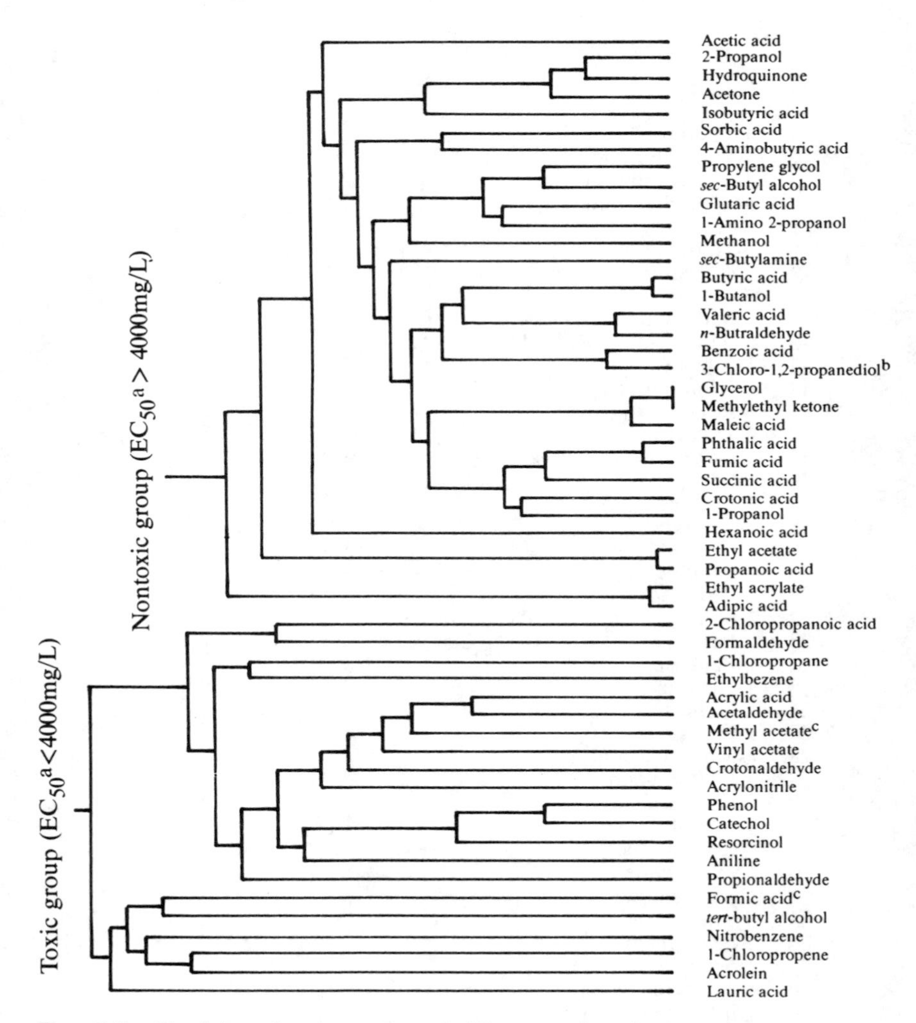

Fig. 2.6. Toxicity dendrogram of 53 petrochemicals developed by cluster analysis. [a] EC_{50} represents the concentration of the chemical required to reduce the activity of methanogenic cultures by 50%; [b] Toxic chemicals misclassified as nontoxic; [c] Nontoxic chemicals misclassified as toxic. Reproduced from Nirmalakhandan and Speece (1988).

As far as predictions of the potential carcinogenic activity of untested chemicals are concerned, Arcos (1987) suggested three categories of criteria within which suspicions might be alerted. These were structural criteria, functional criteria and 'guilt by association' criteria. Structural criteria are based either on formal structural analogies with stabilized types of chemical carcinogen or on considerations of molecular size, shape and symmetry, of electron distribution, of reactive functional groups and their associated steric factors, and of reactive metabolized intermediates, independently of any possible analogy with known chemical carcinogens.

Functional criteria complement structural criteria by representing integrated pharmacological and toxicological capabilities that, irrespective of structure, have shown some degree of correlation with carcinogenic activity. The third class of criteria, 'guilt by association', relates to compounds which, although found inactive according to standard animal bioassays, belong to a chemical class in which several other potent and multitarget compounds are represented. Hay (1988) has discussed complications with some traditional carcinogen bioassay tests.

Among the principal types of known chemical carcinogen that comprise the body of knowledge against which new compounds are evaluated by structural analogy may be listed the following:

1. Aromatic compounds.
2. Azo compounds.
3. Polynuclear hydrocarbons and heteroaromatics.
4. Alkylating agents.
5. Halogenated and polyhalogenated aliphatic and aromatic hydrocarbons and their derivatives.
6. Acylating agents.
7. Miscellaneous epigenetic carcinogens.
8. Carcinogenic metals and metalloids such as cadmium, chromium, nickel and arsenic.
9. 'Foreign-body' carcinogens which either disrupt intercellular homeostasis or mechanically interfere with conformational changes in chromatin or DNA (e.g. asbestos or hard silicates).

Specific examples of QSAR analysis of molecular structures and different modes of exposure affecting transport and toxicities of drugs have been reviewed by Lien (1985). The analyses did not make any *a priori* assumption with respect to mechanisms of absorption, nor did they assume any specific physical structure of the biological membrane. A general equation and its simplified forms were used as follows:

$$\log A = a(\log K)^2 + b \log K + c \log (U/D) + d \log MW + e\chi + f$$

where A is a measure of drug absorption, K is the partition coefficient, (U/D) is the ratio of unionized to ionized species of a drug (for acids, $\log (U/D) = pK_a - pH$); MW is the molecular weight, χ is either the experimental dipole moment or a steric factor, and the coefficients a to f are constants derived from regression analysis. The ideal lipophilic character for maximum absorption ($\log K$) can be obtained by setting the partial first derivative $\log A/\partial \log K$ equal to zero and solving for $\log K_0$. For instance, $\log K_0 = b/2a$.

Using a table, compiled by Rawls (1983), of the minimum lethal toxicities of various poisons, Lien (1985) plotted log 1/MLD (minimum lethal dose) against log molecular weight (Fig. 2.7). The poisons range from sodium cyanide to botulinum toxin A, with a molecular weight range of 10^{13} orders of magnitude. A correlation coefficient of 0·969 was found for the data plotted in Fig. 2.7, with dioxin 'outlying' by more than two standard deviations. This was probably because this MLD was the only one obtained from experiments on guinea-pigs. Bufotoxin and muscarin MLDs were derived from studies on cats, but still fitted the line obtained from the other points for mice data. Deleting dioxin from the regression, Lien (1983) derived the following relationship:

$$\log 1/\text{MLD} = 3{\cdot}067\ (\pm 0{\cdot}411) \log MW - 2{\cdot}028\ (\pm 1{\cdot}509)$$

$$n = 9; \qquad r = 0{\cdot}989; \qquad \text{SD} = 0{\cdot}750$$

Molecular volumes and the toxicity of chemicals to fish have been explored by McGowan and Mellors (1986) with reference to sheepshead minnows *Cyprinodon variecatus* and 96 h LC_{50} exposure. Lipnick *et al.* (1985) have compared favourably toxicity screening data for 55 alcohols with QSAR predictions of minimum toxicity for nonreactive nonelectrolyte organic compounds, and Laughlin *et al.* (1985) found a single linear relationship between a sum of Hansch fragment constants and the 14-day LC_{50} values for zoeae of the mud crab *Rhithropanopeus harrisii* exposed to di- and tri-organotins. In QSARs for nonreactive organic chemicals, Hermens *et al.* (1985) found that hydrophobicity

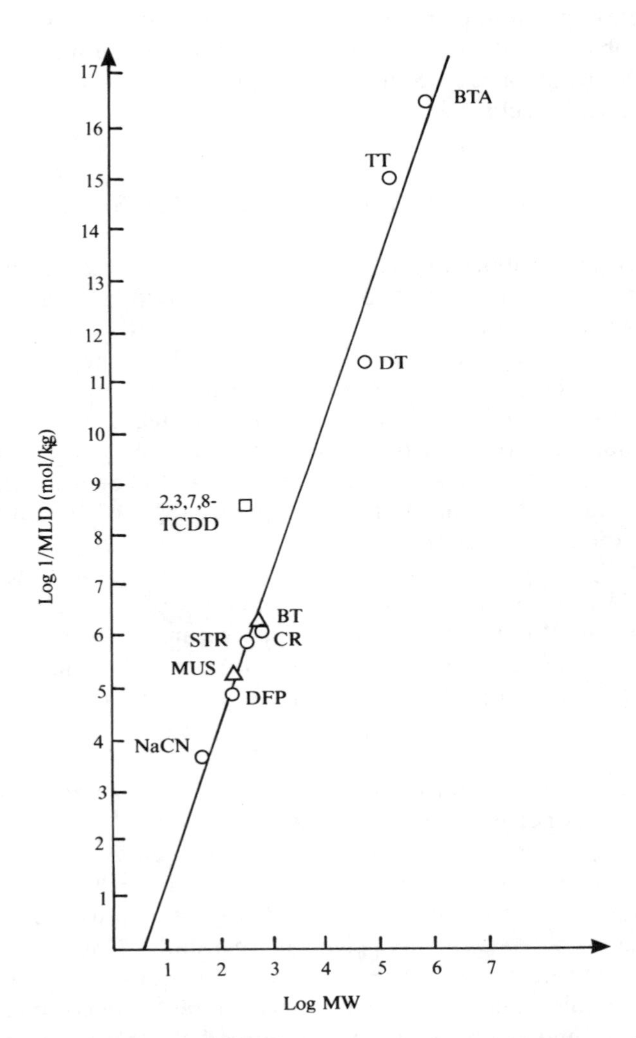

Fig. 2.7. The linear dependence of the minimum lethal dose (MLD) of various poisons on molecular weights. BTA, botulinum toxin A; TT, tetanus toxin; DT, diphtheria toxin; 2,3,7,8-TCCD(dioxin), 2,3,7,8-tetrachlorodibenzo-*p*-dioxin; BT, bufotoxin; CR, curare (the molecular weight of tubocurarine used); STR, strychnine; MUS, muscarin; DFP, di-isopropylfluorophosphate; NaCN, sodium cyanide. Reproduced from Lien (1985).

was the only parameter to give satisfactory results. For anilines, hydrophobicity and the Hammet constant were useful, whereas for reactive organic halogen compounds, aquatic toxicities correlated well with chemical reactivities.

2.3 PARTITION COEFFICIENTS

The partition coefficient (K), considered to be the ratio of the activities of a solute in two immiscible liquid phases, is a key parameter in environmental fate prediction. It provides a clear indication of how a substance will migrate from one environmental phase to another, and is also utilized as a screening tool for assessing potential bioaccumulation in living organisms. Because K is essentially a consequence of molecular structures and charge, the partition coefficient is also correlated with solubility and chemical activity. At very low solute concentrations, it is practicable to substitute molar concentrations for activities.

Generally, the partition law can be treated as an extension to Henry's law (1803), which states that the mass of a gas dissolved by a given volume of liquid at a constant temperature is proportional to the pressure of the gas. Since the concentration of molecules in the gaseous phase is proportional to pressure, it follows that:

$$C_2/C_1 = \text{constant}$$

where C_1 is the concentration of molecules in the gaseous phase and C_2 is the mass per unit volume of gas in solution.

Thus, if a solute is added to two immiscible liquids and is soluble in both, it will distribute itself between them in such a way that its concentration in one solvent is directly proportional to the concentration in the other solvent at a given temperature. This relationship is known as the partition law. For a system such as iodine in benzene (readily soluble) and in water (sparingly soluble), the constant ratio K between the molar concentrations is termed the partition coefficient. This law is only obeyed accurately at low concentrations because it assumes no significant solute–solute and solute–solvent interactions.

Deviations from partition law may occur when molecules dissociate into ions in one solvent but not in the other, as, for example, when hydrogen chloride ionizes in water but not in benzene. Sometimes association rather than dissociation occurs, as when ethanoic acid dimerizes in benzene but not in water, and often solute molecules can exist in different forms in the two phases as is found with ammonia. Smith and White (1929) allowed for both dimerization and ionization in developing equations to describe the distribution of acids between water and toluene, benzene or chloroform.

In addition to problems of ionization and self-association, calculation of true partition coefficients is further complicated by factors such as hydrate formation. In any event, calculations are dependent on the quality of available equilibrium constants. These factors, the thermodynamics of partitioning systems and the energy requirements for phase transfer have been comprehensively reviewed by Leo *et al.* (1971).

The distribution of a solute between two phases in which it is soluble has long been a subject of interest. Realization that the ratio of the concentrations of solute distributed between two immiscible solvents did not depend on the relative volumes of solutions used and was therefore a constant, stemmed from the work of Berthelot and Jungfleisch (1872). Even then it was appreciated that partition coefficient varied with temperature, the more volatile solvent being favoured by a lower temperature.

Nernst (1891) observed that the partition coefficient remained a constant only if a single molecular species was considered partitioned between two phases. Thus, partitioning could be treated as an equilibrium process where the tendency of any single solute molecular species to leave one solvent for another could be a measure of its activity in that solvent and so a function of partial pressure, osmotic pressure and chemical potential. However, since water molecules and solvent molecules will bond with varying degrees of firmness to different solutes, any system more complex than rare gases in hydrocarbons and water becomes impossible to define precisely at the molecular level. For instance, on a molar basis, solvents such as *sec*-butyl alcohol contain more molecules of water in the butanol phase than butanol.

In the early part of this century, Herz (1909) published formulae relating the partition coefficient (K) to the number of extractions necessary to remove a given weight of solute from solution. By the 1950s, the mechanical technique of multiple extraction for separating two closely related solutes had been refined and automated into the procedure known as countercurrent distribution.

Craig and Craig (1950) described countercurrent distribution apparatus as usually consisting of 100 to 200 identical tubes containing a dense solvent. The sample was dissolved in a portion of an upper, lighter solvent and placed in tube 1, which was shaken and allowed to settle. The upper solvent was then automatically transferred to tube 2 while a new portion of upper solvent was added to tube 1. The process was repeated with tubes 2 and 3 and so on. As a result, both solutes in the sample were gradually separated as they were transferred along the sequence of tubes, separation being dependent on their having slightly different partition coefficients in the two immiscible solvents. Partition coefficients may be calculated from the equation:

$$T_{n,r} = \frac{n!}{r!\,(n-r)!}\left(\frac{1}{K+1}\right)^{n(K)^r}$$

where $T_{n,r}$ represents the fraction of the total material in the r tube distributed through n tubes (Williamson & Craig, 1947).

Since partition coefficients are in effect equilibrium constants, it is to be expected that there exist extrathermodynamic relationships between values in different solvent systems. Collander (1951) showed that the equation

$$\log K_2 = a \log K_1 + b \qquad (a, b = \text{constants})$$

held between the systems isobutyl alcohol–water, isopentyl alcohol–water, octanol–water and oleyl alcohol–water. Hansch (1968) further extended the comparison of relatively nonpolar systems using $CHCl_3$–water for K_1 and either CCl_4, xylene, benzene or isoamyl acetate for K_2. Solute–solvent interactions may be studied by defining a reference system and making it the independent variable K_1 in a set of Collander's equations. Mimicking the membrane lipid–water barrier, the octanol–water system consequently has the largest number of measured values and contains the widest selection of functional groups of all the reference systems.

The octanol–water partition coefficient (K_{ow}) is defined as the ratio of a chemical's concentration in the octanol phase to that of its concentration in the aqueous phase of a two-phase octanol–water system, thus;

$$K_{ow} = \frac{\text{concentration in octanol phase}}{\text{concentration in aqueous phase}}$$

Being a ratio, K_{ow} is thus dimensionless. Measured values of K_{ow} for organic chemicals range over 10 orders of magnitude from 10^{-3} to 10^7; consequently, the logarithm of K_{ow} is quoted more commonly, ranging between −3 and 7. Lyman (1982*a*) has reviewed K_{ow} in considerable detail.

When nonpolar phases of partitioning systems contain molecules which cannot hydrogen bond in addition to those which can, Collander's equation gives a regression with a poor correlation coefficient and high standard deviation (Smith, 1921). Leo *et al.* (1971) suggested that solute types could easily be distinguished by separating 'minus deviants' and 'plus deviants' from a collection of values for a single equation. Combining the data for several solvent systems, it then becomes apparent that the strong hydrogen bond donors are 'minus deviants' and the hydrogen bond acceptors are 'plus deviants'. The ether–water system is unusual, for while it segregates donors from acceptors, deviations are reversed.

Table 2.5 presents a classification of solutes as hydrogen donors

Table 2.5
General solute classes

Group 'A' H donors	1.	Acids
	2.	Phenols
	3.	Barbiturates[a]
	4.	Alcohols[a]
	5.	Amides (negatively substituted, but not di-*N*-substituted)[a]
	6.	Sulphonamides[a]
	7.	Nitriles[a]
	8.	Imides[a]
	9.	Amides[a,b]
Group 'B' H acceptors	10.	Aromatic amines (not di-*N*-substituted)[a,b]
	11.	Miscellaneous acceptors
	12.	Aromatic hydrocarbons
	13.	Intramolecular H bonds[c]
	14.	Ethers
	15.	Esters
	16.	Ketones
	17.	Aliphatic amines and imines
	18.	Tertiary amines (including ring N compounds)

[a] 'Neutral' in $CHCl_2$ and CCl_4.
[b] Classes 9 and 10 must be reversed when considering the ether and oil solvent systems.
[c] For example, *o*-nitrophenol.
Reproduced from Leo *et al.* (1971).

(Group A) and hydrogen acceptors (Group B). If a solute molecule contained two or more noninteracting functional groups each of which required an A or B classification, Leo *et al.* (1971) placed it according to its closeness of fit to the solvent regression equation describing the group.

The slopes and intercepts of these equations, which take the form:

$$\log K_{\text{solvent}} = a \log K_{\text{octanol}} + b$$

may be used to study solute–solvent interactions as compared with the reference solute–octanol interaction. In Table 2.6, which orders solutes according to the amount of water they contain at saturation, the slope provides a measure of the solvent system's sensitivity to changes in solute lipophilicity. Butanol–water has the lowest slope value and also the least sensitivity. Thus, when each of this pair is saturated with the other, phase differences are negligible. Increasing the hydrocarbon chain length in the alcohol solvents increases the dissimilarity of the alcohol–water phase, and sensitivity to solute changes increases. A maximum sensitivity is apparently reached at octanol (slope = 1·000), for the slope in the oleyl alcohol equation is unchanged (0·999).

The intercept value in the regression equation for a given solvent system in Table 2.6 is $\log K$ for any solute that is distributed equally between water and octanol. Thus, $\log K_{\text{octanol}}$ equals zero. A negative intercept indicates that the solvent is more lipophilic than octanol, and a positive intercept that it is more hydrophilic. Leo *et al.* (1971) used the intercept values from the 'A' or 'sole' equation as a measure of the solvent's lipophilicity, and found a good correlation between these values and the water content at saturation:

$$\log (H_2O) = 1.077\,[\text{intercept}] + 0{\cdot}249$$

where n equals 17; r is 0·979 and s is 0·217.

This equation shows that the inability of a partition solvent to 'accommodate' water provides a measure of its lipophilic behaviour towards a wide variety of organic solutes.

2.3.1 Additive–constitutive properties

Meyer (1899) and Overton (1901) and others first recognized that in a homologous series the partition coefficient increased by a factor of between two and four per CH_2 group. Later, Fujita *et al.* (1964) established additivity for a variety of groups, defining the substituent

Table 2.6
Solvent regression equations

H_2O concn. at saturation		*$\log K_{solv} = a \log K_{octanol} + b$*									
Solvent (vs. H_2O)	$10^3 M$	*H-donor solutes (equation 'A')*					*H-acceptor solutes (equation 'B')*				
		a	*b*	*n*	*r*	*s*	*a*	*b*	*n*	*r*	*−s*
Cyclohexane	2·5	0·675	− 1·842	*26*	0·761	0·505	1·063	− 0·734	*30*	0·957	0·360
Heptane	3·3	1·056	− 2·851	*10*	0·764	0·916	1·848	− 2·223	*11*	0·954	0·534
CCl_4[a]	10·0	1·168	− 2·163	*24*	0·974	0·282	1·207	− 0·219	*11*	0·959	0·347
Xylene	18·8	0·942	− 1·694	*19*	0·963	0·225	1·027	− 0·595	*21*	0·986	0·230
Toluene	25·6	1·135	− 1·777	*22*	0·980	0·194	1·398	− 0·922	*14*	0·971	0·274
Benzene	26·0	1·015	− 1·402	*33*	0·962	0·234	1·223	− 0·573	*19*	0·958	0·291
$CHCl_3$[b]	68·4	1·126	− 1·343	*28*	0·967	0·308	1·276	+ 0·171	*21*	0·976	0·251
Oils[c]	72·5	1·099	− 1·310	*65*	0·981	0·271	1·119	− 0·325	*14*	0·988	0·233
Nitrobenzene	180	1·176	− 1·072	*9*	0·977	0·217					
Isopentyl acetate	456	1·027	+ 0·072	*22*	0·986	0·209					
Ether	690	1·130	− 0·170	*71*	0·988	0·186	1·142	− 1·070	32	0·957	0·326

(continued)

Table 2.6—contd.

H_2O concn. at saturation		*$\log K_{solv} = a \log K_{octanol} + b$*				
Solvent (vs. H_2O)	$10^3 M$	*H-donor solutes (equation 'A')*				
		a^a	b^a	n	r	s
		'Sole' equation				
Oleyl alcohol	712	0·999	− 0·575	*37*	0·985	0·225
MIBK	950	1·094	+ 0·050	*17*	0·993	0·184
Ethyl acetate	1 620	0·932	+ 0·052	*9*	0·969	0·202
Octanol	2 300	1·000	+ 0·000			
Cyclohexane	4 490	1·035	+ 0·896	*10*	0·972	0·340
1° Pentanols	5 000[d]	0·808	+ 0·271	*19*	0·987	0·161
2° and 3° Pentanols	5 320[e]	0·892	+ 0·288	*11*	0·996	0·091
2-Butanone	5 460	0·493	+ 0·315	*9*	0·987	0·093
Cyclohexanol	6 510	0·745	+ 0·866	*12*	0·985	0·100
1° Butanols	9 440[f]	0·697	+ 0·381	*57*	0·993	0·123

MIBK: Methylisobutylketone.
[a] The 'N' equation is $\log K_{CCl_4} = 0{\cdot}862 \log K_{octanol} - 0{\cdot}626$ ($n = 6$, $r = 0{\cdot}809$, $s = 0{\cdot}462$).
[b] The 'N' equation is $\log K_{CHCl_3} = 1{\cdot}10 \log K_{octanol} - 0{\cdot}617$ ($n = 32$, $r = 0{\cdot}974$, $s = 0{\cdot}254$).
[c] Most liquid glyceryl triesters fit this equation; olive, cottonseed and peanut oils were the most frequently used.
[d] *n*-Amyl alcohol = 5·03 M in water; isoamyl alcohol = 4·50 M in water.
[e] Water content measured for 2-pentanol only.
[f] Water content measured for 1-butanol only.
K values for all of the solutes used to develop the equations are listed in *J. Org. Chem.* **36** (1971) 1539 (microfilm).
Data from Leo *et al.* (1971).

constant π as

$$\pi_x = \log K_x - \log K_H$$

where K_H is the parent molecule, K_x is a derivative of K_H and π_x is the logarithm of the partition coefficient of the functional group x. For instance, π_{Cl} can be obtained as follows:

$$\pi_{Cl} = \log K_{chlorobenzene} - \log K_{benzene}$$

Table 2.7 contains a partial list of some of the more common functional groups. For more complex structures, intensive searching may be required in order to find a suitable value for the parent

Table 2.7
List of π values for common functional groups

Function	*Aromatic*	*Aliphatic*
NH_2	−1·23	−1·19
I	1·12	1·00
$S—CH_3$	0·61	0·45
$COCH_3$	−0·55	−0·71
$CONH_2$	−1·49	−1·71
$COOCH_3$	−0·01	−0·27
Br	0·86 (*ortho*)	0·60
Br	1·12 (*para*)	
CN	−0·57	−0.84
F	0·14	−0.17
Cl	0·76 (*ortho*)	0·39
Cl	1·04 (*meta*)	
Cl	0·98 (*para*)	
COOH	−0·28	−0·67
OCH_3	−0·02	−0·47
OC_6H_5	2·08	1·61
$N(CH_3)_2$	0·18	−0·30
NO_2	−0·28	−0·85
CH_3	0·84 (*ortho*)	
CH_3	0·60 (*para*)	
CH_3	0·54 (*meta*)	
OH	−0·31 (*para*)	−1·16
C_6H_6	2·13	
C_6H_5OH	1·46	
$—CH_2—$		0·50
CF_3	1·07 (*para*)	

Data from Neely (1980).

structure. Once the K value has been assigned, a reasonably good estimate of many different types of derivatives can be calculated. Leo *et al.* (1971) and Hansch *et al.* (1972) indicated some of the uncertainties in this type of calculation. The following two calculations for *m*-chlorotoluene and chlorpyrifos (Neely, 1980) serve as illustrative examples of the calculation of partition coefficient according to the scheme:

$$\log K_x = \sum_i \pi_{x_i} + \log K_H$$

where the summation is for all the substituents x_i on the parent structure.

log K for *m*-chlorotoluene

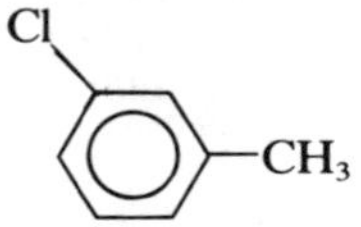

		Calculated	Experimental
1.	log K_H (benzene)	2·13	
2.	π_{Cl}	0·76	
3.	π_{mCH_3}	0·54	
4.	log K_x	3·43	3·28

log P for chlorpyrifos

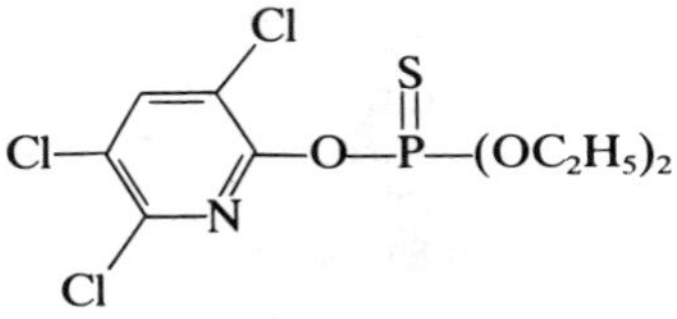

		Calculated	Experimental
1.	log K for  (benzene ring)	3·46	
2.	log K for 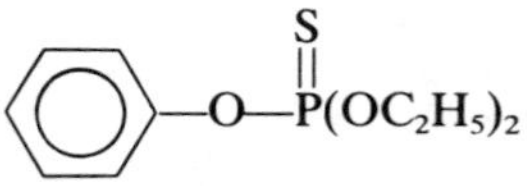 (phenyl)—O—P(S)$(OC_2H_5)_2$	2·13	
3.	log K for —O—P(S)$(OC_2H_5)_2$ = (1–2)	1·33	
4.	log K for pyridine 	0·66	
5.	log K for Cl-(*ortho*) 2 × log K for Cl-(*meta*)	0·76 2·08	
6.	log K (chlorpyrifos) = (3 + 4 + 5)	4·83	4·81

2.3.2 Fragment constants

A fragment is an atom or string of atoms whose exterior bonds are to isolating carbon atoms. (An isolating carbon is one that either has four single bonds, at least two of which are to nonheteroatoms, or is multiply bonded to other carbon atoms (Lyman, 1982).)

Often a K value for the parent compound is not easily obtainable. Leo (1975) therefore proposed that the partition coefficient be estimated by the addition of a series of fragment constants according to

$$\log K = \sum_{1}^{n} a_n f_n$$

where f_n is the fragment constant and a_n is the number of fragments.

As such an equation falsely implies that the hydrophobicity of any structural fragment is invariant, rules governing the attachment of different fragments have been developed to account for the observed variation (Leo, 1975). Unfortunately, such rules became as long and

Table 2.8
Fragment constants to be used for calculating partition coefficients

Fragment	*Constant*		
	Aliphatic	*Aromatic ring*	*Double aromatic ring*
—CH	0·89		
—CH_2—	0·66		
—CH	0·43		
C	0·20		
—H	0·23		
—Br	0·20	1·09	
—Cl	0·06	0·94	
—F	−0·38	0·37	
—N	−2·16	−1·17	−1·29
—NO_2	−1·26	−0·02	
—O—	−1·81	−0·57	0·53
—S—	−0·79	0·03	0·77
—NH—	−2·11	−1·03	−0·18
—NH_2	−1·54	−1·00	
—OH	−1·64	−0·40	
—CN	−1·28	−0·34	
—C_6H_5	1·90		

Data from Neely (1980).

involved as the structure was complicated. Among the rules elucidated for branching and chain formation are the following:

A. Every single bond after the first makes a negative contribution to hydrophobicity. This has a value of −0·09 for cyclic compounds.
B. A value of −0·13 for every branch is added, and −0·22 is used for a group branch.

These two rules and the fragment constants presented in Table 2.8 are illustrated as follows by four examples of the use of fragment constants in the estimation of partition coefficients (Overton, 1901; Fujita *et al.*, 1964).

Four examples of the use of fragment constants in the estimation of partition coefficients (Neely, 1980)

	log K	
	Calculated	*Observed*
1. Cyclopropane (triangle)		
$3f_{CH_2} + 2f_b$ $3(0{\cdot}66) - 2(0{\cdot}09)$	1·80	1·72
2. Cyclohexane (hexagon)		
$6f_{CH_2} + 5f_b$ $6(0{\cdot}66) - 5(0{\cdot}09)$	3·51	3·44
3. Chain branching $CH_3CH(CH_3)CH_3$ (H above, CH_3 below central C)		
$3f_{CH_3} + f_{CH} + 2f_b + f_{cbr}$ where f_{cbr} is the chain branching constant (−0·13)	2·73	2·76
4. Group branching HO—$CH(CH_3)_2$		
$2f_{CH_3} + f_{CH} + f_{OH} + 2f_b + f_{gbr}$ where f_{gbr} is the group branching constant (−0·22)	0·11	0·05

2.3.3 Molecular connectivity

The connectivity index (χ) has been discussed by Kier *et al.* (1975), Hall *et al.* (1975) and Kier *et al.* (1976). Its relationship with several chemical and physical properties has been detailed in Section 2.2.5. Murray *et al.* (1975) found a significant correlation ($r = 0{\cdot}986$, S.E. = $0{\cdot}152$) between χ and $\log K$ for 138 substances, including 24 esters, 9 carboxylic acids, 49 alcohols, 16 ketones and 12 ethers.

2.3.4 Experimental determination of partition coefficient

The most extensive and useful partition coefficient data have been obtained by simply shaking a solute with two immiscible solvents and then analysing the solute concentration in one or both phases (Leo *et al.*, 1971). Such experiments were performed by Campbell *et al.* (1983) in order to evaluate partition coefficients for aniline, naphthalene, naphthol, phenol and pyridine between water and several organic solvents. With some modification to the 'shake flask' method, Marple *et al.* (1986) measured K_{ow} for 2,3,7,8-tetrachlorodibenzo-*p*-dioxin. However, a variety of different approaches and methods have proved successful and some of these are listed in Table 2.9. Brooke *et al.* (1986) have discussed the measurement and estimation of octanol/water partition coefficients with particular reference to high values ($\log k > 5$).

2.3.5 Activity coefficients

The partition coefficient may also be expressed in terms of solute activity coefficients since

$$K_{ow} = 0{\cdot}151\gamma_c^w/\gamma_c^o$$

where K_{ow} is the octanol–water partition coefficient; γ_c^o is the activity coefficient of the chemical (c) in the octanol phase and γ_c^w is the activity of the chemical in the water phase.

The activity coefficient and the mole fraction of a chemical in equilibrium between the phases of an octanol/water system are related as follows:

$$X_c^o\gamma_c^o = X_c^w\gamma_c^w$$

therefore

$$X_c^oX_c^w = \gamma_c^w/\gamma_c^o$$

where X_c^o is the mole fraction of chemical (c) in the octanol (o) phase and X_c^w is the mole fraction of chemical (c) in the water (w) phase.

Table 2.9
Methods for determining partition coefficients

Technique	*Reference*
Ratio of solubilities in two separate solvents (K at saturation only)	Wroth and Reid (1916)
Countercurrent distribution	Craig *et al.* (1947)
Volatile solute concentration in two miscible or immiscible solvents	Christian *et al.* (1963)
Automatic titration for K of organic bases between immiscible solvents	Brandstrom (1963)
Diffusion techniques for solute with surfactant properties	Greenwald *et al.* (1961)
Reverse-phase thin-layer chromatography	Ellgehausen *et al.* (1981) Bruggeman *et al.* (1982)
Reverse-phase high-performance liquid chromatography	Carlson *et al.* (1975) McCall (1975) Veith *et al.* (1979) McDuffie (1981)
Generator column high-performance liquid chromatography	Woodburn *et al.* (1984) Doucette and Andren (1987)
Equilibriated water-phase sampling	Marple *et al.* (1986)

Regular solution theory models (Weimer & Prausnitz, 1965; Helpinstill & Van Winkle, 1968) and linear free energy relationships (Leo & Hansch, 1971) as a means of estimating aromatic solute partition coefficients have been considered by Campbell *et al.* (1983) and Campbell and Luthy (1985). The latter authors used UNIFAC, a group contribution model capable of predicting activity coefficients for nonelectrolytes in liquid mixtures (Fredenslund *et al.*, 1975). Banerjee (1984, 1985) and Banerjee and Howard (1988) have further discussed the estimation of solubility and partitioning of organic compounds in water with UNIFAC. Also used was the CLOGP algorithm for K_{ow} (Leo & Hansch, 1986), which is accurate for compounds of log K_{ow} of up to about five. Banerjee and Howard (1988) found that K_{ow} ranging over nine orders of magnitude for 75 compounds could be correlated ($r = 0{\cdot}98$) with UNIFAC-derived activity coefficients when

$$\log K_{ow} = -0{\cdot}40 + 0{\cdot}73 \log (\gamma_w)_u - 0{\cdot}39 \log (\gamma_o)_u$$

where $(\gamma_w)_u$ and $(\gamma_o)_u$ are the activity coefficients in water and octanol, respectively. The constants 0·73 and −0·39 were obtained empirically and are intended to compensate for group nonadditivity.

2.4 PARTITION COEFFICIENT AND BIOCONCENTRATION

Bioconcentration and bioaccumulation are essential links in the sequential increase in concentration of a chemical when transferring from one trophic level to the next, which occurs in the process known as biomagnification. Bioconcentration may be considered to have occurred when an organism accumulates an environmental substance to concentrations above those normally associated with that substance in the immediate environment. The bioconcentration factor (BCF), i.e. the ratio of a chemical's concentration in the organism (wet weight) to that in the ambient environment at equilibrium, provides a useful measure of the bioconcentration potential of a chemical. Values of BCF range from about one to over one million. Esser and Moser (1982) cited three options for the determination of BCF:

(a) Deduction from the partition coefficient, *n*-octanol–water.
(b) Determination of steady-state partitioning.
(c) Determination of the ratio of uptake and depuration rates.

Table 2.10
Factors relating to bioconcentration testing

Structural transformations	Chemical stability Photodegradability Biodegradability	Selection of bioconcentration test system ↓
Physicochemical properties	S_{water} K_{ow} (calculated) K_{oc} C_{water}/C_{air} pK MW	K_{ow} ↓ Static fish test ↓ Dynamic fish test
Analytical procedures	Parent Impurities Polar degradation products	

After Esser and Moser (1982).

Table 2.10 outlines some factors involved in the determination of bioconcentration potential. In aquatic test systems, factors contributing to BCF variation include the animal (e.g. fish) species, age and feeding regimen, the experimental conditions, analytical accuracy and precision, and properties of the test chemical such as solubility in water, sorptivity, ionization, molecular size and degradability.

The instantaneous rates of uptake and BCFs of nine organic compounds were determined by Mailhot (1987) utilizing radioactively labelled compounds in the green alga *Selenastrum capricornutum*. The specific activities and 10 physicochemical properties of the nine organic compounds studied are presented in Table 2.11. Seven are hydrocarbons, one a triazine (atrazine) and one an acid (benzoic acid). Molar volume is the volume occupied by one mole and is numerically equal to the molecular weight divided by the density. Parachor is a molecular size parameter best conceptualized as molar volume corrected for surface tension. The capacity ratio (k'), in this case determined experimentally by reverse-phase high performance liquid chromatography, was calculated as follows:

$$(T_r - T_m)/T_m$$

where T_r and T_m were the retention times of the sample solute and the mobile phase (KNO_3), respectively.

Also given in Table 2.11 are the BCFs and the instantaneous uptake rates for the nine compounds. Regression of these values against each of the physicochemical properties identified capacity ratio and K_{ow} as parameters which best and equally well predicted biocentration factor. If, however, the hydrocarbons alone were considered, then $\log k'$ predicted BCF most accurately. A linear plot of log (BCF) against $\log k'$ or $\log K_{ow}$ for all nine compounds could be achieved by a power transformation of BCF where

$$\text{BCF}^{0{\cdot}4} = 6{\cdot}9 + 28 \log k'$$

and

$$\text{BCF}^{0{\cdot}4} = -21 + 12 \log K_{ow}$$

If only the five most lipophilic compounds were considered, excluding atrazine, benzoic acid, chloroform and benzene hexachloride, then log (BCF) could be described in the following manner:

$$\log(\text{BCF}) = 3{\cdot}1 + 0{\cdot}74 \log k'$$

and

$$\log(\text{BCF}) = 2{\cdot}6 + 0{\cdot}28 \log K_{ow}$$

Table 2.11
Specific activity and value of 10 physicochemical properties for algal BCF and uptake rate for nine organic compounds

Compound	*SA* ($Ci\ mol^{-1}$)	*log* K_{ow}	*log* k^1	*log S* (*M*)	(χ)	*Para.*	*MV* (cm^3)	*MW*	*MP* (°C)	*D* ($g\ cm^{-3}$)	*BP* (°C)	*BCF*	*UR* ($\%\ min^{-1}$)
Benzoic acid	29·4	1·87	−0·24	−1·65	3·01	175	92·4	122	122	1·32	249	7·6	—
Atrazine	25·0	2·69	0·16	−3·84	5·71	456	—	216	174	—	—	20	—
Chloroform	13·9	1·94	−0·10	−1·15	1·73	190	80	119	−64	1·49	62	690	0·09
Benzene hexachloride	62·0	4·06	0·62	−4·70	5·46	479	156	291	112	1·87	323	1 500	—
Tetrachlorobenzene	7·2	4·60	1·06	−5·08	3·65	364	116	216	139	1·86	245	7 700	0·03
Anthracene	15·1	4·49	1·01	−6·45	4·81	410	143	178	216	1·25	340	7 800	0·11
DDT	40·0	5·70	1·58	−7·68	7·01	663	228	353	109	1·55	—	12 000	0·44
Hexachlorobenzene	13·5	5·64	1·66	−7·36	4·50	443	182	285	229	1·57	324	26 000	0·09
Hexachlorobiphenyl	12·5	7·08	1·98	−7·98	6·56	618	232	361	103	1·56	—	37 000	—

SA: Specific activity; log K_{ow}: log *n*-octanol–water partition coefficient; log k^1: log capacity ratio; log *S*: solubility; χ: molecular connectivity index; Para: parachor; MV: molecular volume; MP: melting point; D: density; BP: boiling point; BCF: bioconcentration factor in the green alga *Selenastrum capricornutum*; UR: uptake rate.
Data from Mailhot (1987).

The uptake rate, which could only be measured effectively for five compounds was best predicted by the molecular connectivity index. Subsequently, Mailhot and Peters (1988) used regression analyses to relate partition coefficients to nine physicochemical properties, using data for 301 organic compounds from 10 chemical families. Solubility was found to be the property predicting the partition coefficient most precisely.

Neely *et al.* (1974) determined the bioconcentration of several chemicals in trout (*Salmo gairdnerii*) muscle, and found that a linear relationship existed between BCF and the *n*-octanol–water partition coefficient of the chemical. The equation for the line of best fit for the chemicals given in Table 2.12 was as follows:

$$\log(\mathrm{BCF}) = 0{\cdot}542 \log K + 0{\cdot}124$$

The linear regression yielded a multiple correlation coefficient of 0·948. An F-test indicated a 0·999 confidence level. The equation also successfully predicted the bioconcentration of other chemicals from their partition coefficients.

Mackay (1982) extended the one-constant correlation derived for fish BCF and octanol–water K to describe a simple relationship between BCF and aqueous solubility, while Chiou (1985) has determined the triolein–water partition coefficients (K_{tw}) for 38 slightly water-soluble organic compounds. (Lipid triolein ($C_{57}H_{104}O_6$) is also

Table 2.12
Partition coefficient (K) and bioconcentration factor (k_1/k_2) for several chemicals in trout muscle

Chemical	*log K*[a]	*log* (k_1/k_2)[a]
1. 1,1,2,2-Tetrachloroethylene	2·88	1·59
2. Carbon tetrachloride	2·64	1·24
3. *p*-Dichlorobenzene	3·38	2·38
4. Diphenyl oxide	4·20	2·29
5. Diphenyl	4·09	2·64
6. 2-Biphenyl phenyl ether	5·55	2·74
7. Hexachlorobenzene	6·18[b]	3·89
8. 2,2′,4,4′-Tetrachlorodiphenyloxide	7·62[b]	4·09

[a] These are given in log to the base 10; k_1 = uptake rate (h^{-1}) and k_2 = clearance rate (h^{-1}).
[b] Calculated according to Leo *et al.* (1971).
Data from Neely *et al.* (1974).

known as glyceryl trioleate.) When published BCF values of organic compounds in fish were based on the lipid content rather than the total mass, they were found to be approximately equal to the K_{tw}. This suggests near equilibrium for solute partitioning between water and fish lipid, and is consistent with suggestions that the lipid content of the fish is the principal component for concentrating (nonionic) organic compounds.

Pereira *et al.* (1988) found that bioconcentration factors of halogenated organic compounds determined on the basis of lipid content correlated reasonably well with equilibrium K_{tw} partition coefficient for Atlantic croakers (*Micropogonias undulatus*), blue crabs (*Callinectes sapidus*), spotted sea trout (*Cynoscion nebulosis*) and blue catfish (*Ichtalurus furcatus*).

Using BCF data from guppies (Könemann & van Leeuwen, 1980) and rainbow trout (Oliver & Niimi, 1983) a linear plot of log BCF against log K_{tw} was recorded. The correlation ($n = 18$, $r^2 = 0{\cdot}915$) gave the following equation:

$$\log(\text{BCF}) = 0{\cdot}957 \log K_{tw} + 0{\cdot}245$$

When log (BCF) values of the same compound were correlated with their corresponding log K_{ow} values a squared correlation coefficient of 0·904 resulted equating:

$$\log(\text{BCF}) = 0{\cdot}893 \log K_{ow} + 0{\cdot}607$$

In general, it was found that for most substituted benzenes of comparable molar volume, log K_{tw} could be satisfactorily correlated with log K_{ow}. A linear regression ($n = 25$, $r^2 = 0{\cdot}995$) between log K_{tw} and log K_{ow} gave the equation:

$$\log K_{tw} = 1{\cdot}00 \log K_{ow} = 0{\cdot}105$$

Bioconcentration factor in aquatic organisms has been considered in some detail by Bysshe (1982) who warned that many compounds (log (BCF) > 6) move across membranes very slowly, thus BCF measurements may be relatively low until 20 to 30 days equilibrium has elapsed. The relationship between bioconcentration of hydrophobic chemicals in fish and membrane permeation has been discussed by Gobas *et al.* (1986).

A quantitative relationship has also been found to exist between the lipophilicity, expressed as K_{ow}, of a wide range of organic compounds and their BCF in human adipose tissue (Geyer *et al.*, 1987). Figure 2.8

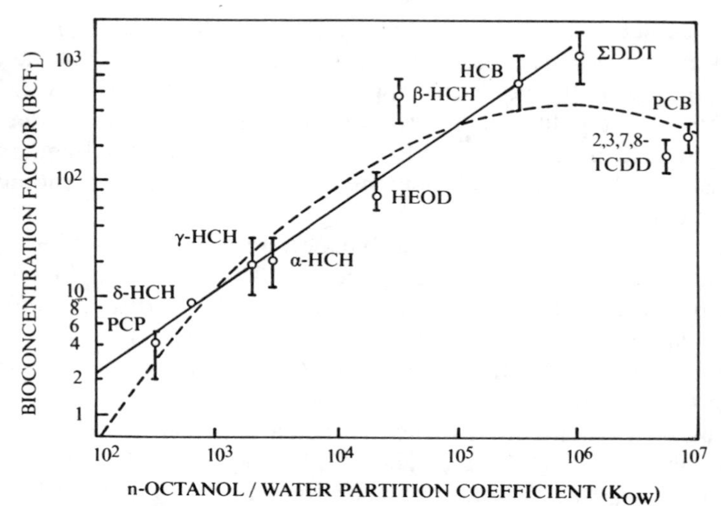

Fig. 2.8. Relationship between log bioconcentration factor, BCF_L (on a lipid weight basis) of organic environmental chemicals in human adipose tissue and log *n*-octanol/water partition coefficient. Each point represents the mean and range of BCF_L. Solid line = eqn (2), dashed line = eqn (4). PCP, pentachlorophenol; α-HCH, α-hexachlorocyclohexane; β-HCH, β-hexachlorocyclohexane; δ-HCH, δ-hexachlorocyclohexane; γ-HCH, lindane; HEOD, dieldrin; HCB, hexachlorobenzene; ΣDDT, DDT + DDE + DDD; 2,3,7,8-TCDD, 2,3,7,8-tetrachlorodibenzo-*p*-dioxin; PCB, polychlorinated biphenyls. Reproduced from Geyer *et al.* (1987).

illustrates the relationships that could be equated either linearly as

$$\underset{\text{(wet weight)}}{\log(\mathrm{BCF})} = 0{\cdot}756 \underset{\text{(lipid)}}{\log K_{\mathrm{ow}}} - 1{\cdot}415 \qquad (6)$$

or

$$\underset{\text{(lipid)}}{\log(\mathrm{BCF})} = 0{\cdot}745 \underset{\text{(lipid)}}{\log K_{\mathrm{ow}}} - 1{\cdot}19 \qquad (7)$$

and using the parabolic model of Hansch (1969), widely used to describe mathematically the absorption of drugs, PCBs and tetrachlorodibenzodioxin, are included by

$$\underset{\text{(wet weight)}}{\log(\mathrm{BCF})} = 2{\cdot}56 \underset{\text{(lipid)}}{\log K_{\mathrm{ow}}} - 0{\cdot}22 (\log K_{\mathrm{ow}})^2 - 4{\cdot}82 \qquad (8)$$

or

$$\underset{\text{(lipid)}}{\log(\mathrm{BCF})} = 2{\cdot}54 \underset{\text{(lipid)}}{\log K_{\mathrm{ow}}} - 0{\cdot}22 (\log K_{\mathrm{ow}})^2 - 4{\cdot}56 \qquad (9)$$

Such equations allow for the estimation of the likely maximum BCF of a new chemical in humans if the K_{ow} is known.

However, the K_{ow} is not without its critics when considered to be a good guide to biconcentration factor. Schüürmann and Klein (1988) reexamined the log–log relationships of BCF with different types of K_{ow} data for a set of 49 diverse compounds and found only a moderate overall correlation between log (BCF) and log K_{ow}. The correlation improved substantially when the compounds considered were restricted to the subsets of chlorinated hydrocarbons (CHCs) and PAHs. It was found that for combined CHCs and PAHs, the solvent accessible surface area (SASA) and molar refraction were better predictors of log (BCF) than log K_{ow} values, calculated according to Leo's CLOGP-3·42 computer program.

Table 2.13 illustrates that empirically a large number of linear regression relationships between log (BCF) and log K_{ow} have been

Table 2.13
Linear regression relationships between log (BCF) and log K_{ow}

Author(s)	*Regression equations*[a]			
	a	*b*	*n*	*r*
Neely *et al.* (1974)	0·54	0·12	8	0·95
Lu and Metcalf (1975)	0·63	0·73	11	0·79
Metcalf *et al.* (1975)	1·16	0·75	9	0·98
Veith *et al.* (1979*b*)[b]	0·85	−0·70	55	0·95
Kenaga and Goring (1980)	0·77	−0·97	36	0·76
	0·94	−1·50	26	0·87
Könemann and van Leeuwen (1980)[c]	0·98	−0·06	6	0·99
Ellgehausen *et al.* (1980)	0·83	−1·71	8	0·98
Veith *et al.* (1980*a*)	0·46	0·63	25	0·63
Veith *et al.* (1980*b*)	0·76	−0·23	84	0·82
Saarikoski and Viluksela (1982)	1·02	−1·82	9	0·98
Mackay (1982)	1·00	−1·32	44 (?)	0·95
Veith and Kosian (1983)	0·79	−0·40	122	0·93
Oliver and Niimi (1983)	1·02	−0·63	11	0·99
Davies and Dobbs (1984)	0·60	0·19	31	0·75
Chiou (1985)	0·89	0·61	18	0·95
Zaroogian *et al.* (1985)	0·61	0·69	11	0·84
Connell and Schüürmann (1988)	0·94	−1·10	49	0·89
Schüürmann and Klein (1988)	0·78	−0·35	22	0·95

[a] $\log(\mathrm{BCF}) = a \log K_{ow} + b$.
[b] log K_{ow} estimated chromatographically.
[c] Based on lipid weight content.

derived with the general form:

$$\log(\mathrm{BCF}) = a \log K_{ow} + b$$

It can be seen that in most cases the slope (a) was found to be less than unity, and for the intercept (b) there was a majority of negative values according to theoretical interpretations.

Chapter 3

Chemical Release and Environmental Pathways

3.1 NATURAL PHENOMENA

A number of natural phenomena contribute significantly to environmental concentrations of polluting substances on a global scale. For example, dusts may be windblown many thousands of miles from their point of origin. Forest, bush and grass fires caused by lightning release large quantities of smoke and trace gases into the atmosphere. Forests also release volatile hydrocarbons (terpenes) which, during sunny weather conditions, may participate in photochemical reactions. Substantial releases of sulphur dioxide and particulates are associated with volcanic emissions. Because some plumes penetrate the stratosphere, the atmospheric residence times of such substances may be long and in consequence the atmospheric radiation balance can be altered and world climate influenced (Whelpdale & Munn, 1976). Sea spray is a major source of atmospheric particulate matter and trace gases are sometimes transferred from the oceans to the atmosphere by evaporation. At times, however, adsorption from the atmosphere is the dominant process. For example, the oceans generally act as a sink for carbon dioxide, although during periods of intense phytoplankton blooming they may act as a source.

The actual size of a potential receptor for contaminants can in some instances be used as a guide to its susceptibility to anthropogenic contamination, without any information as to specific pollutants. The Pollution Potential Index (PPI) developed by Zoeteman (1973) is a planning index based on the size of the population within a given drainage area, the degree of economic activity (measured by the *per capita* Gross National Product), and the average flow rate of the river:

$$\mathrm{PPI} = \frac{NG}{Q} \times 10^6$$

where N is the number of people living in a drainage area, G is the average *per capita* Gross National Product and Q is the yearly average flow rate ($m^3\,s^{-1}$).

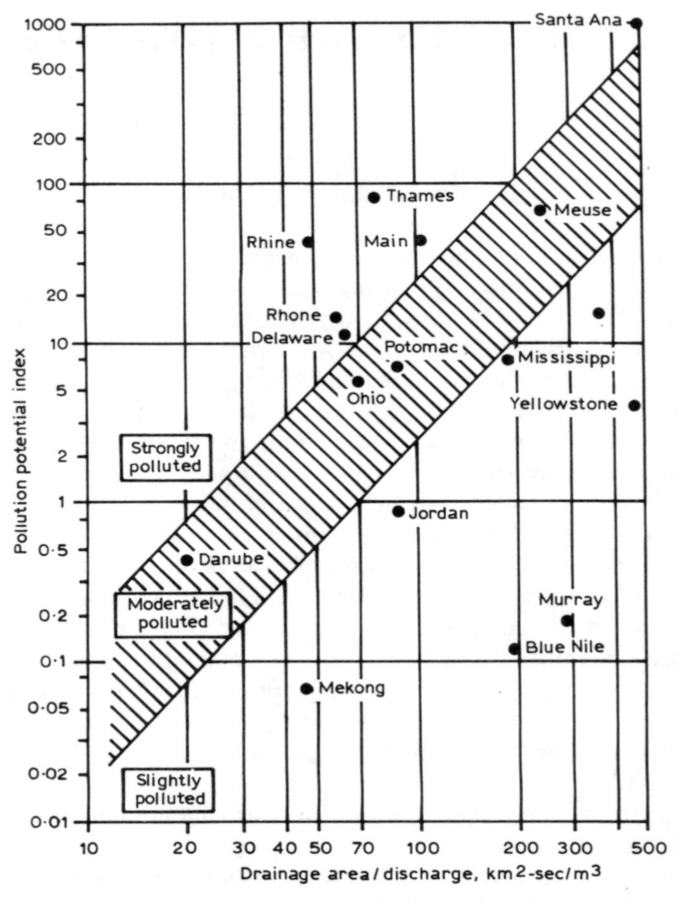

Fig. 3.1. A classification of the world's rivers with respect to their potential for artificial pollution. Reproduced from Ott (1978) (after Zoeteman (1973)).

By considering the ratio of the drainage area of a river to its flow rate, Zoeteman (1973) classified rivers into three groups according to their potential for artificial contamination. As illustrated in Fig. 3.1, these categories were 'slightly polluted', 'moderately polluted' and 'strongly polluted'. Although very simple, and ignoring the possible impact of pollution abatement policies, the PPI does stimulate thought about water pollution problems. Other environmental indices have been most comprehensively reviewed by Ott (1978).

3.2 HUMAN ACTIVITIES

In the countryside, agricultural mismanagement often contributes to soil erosion and so exacerbates problems of windblown dust. Slash burning and decaying farm wastes release a wide variety of substances into the atmosphere. Pesticide and fertilizer applications also add to the anthropogenic release of chemicals into the environment. Figure 3.2 illustrates some distinct human activities influencing the environmental cycling of chemicals and determining the medium to which they are initially released.

Accidents

The occurrence of accidents are naturally difficult to predict; however, it is possible to be prepared for a given worst-case scenario. The cost-benefit problem of how much expenditure is justified to guard against a given probability of occurrence is a separate issue (Go, 1988). In fate prediction studies, one assumes that fail-safe mechanisms have not succeeded in their design objectives.

There are numerous tragic examples of such a situation. Methyl isocyanate (MIC) is a highly toxic chemical that was responsible for the deaths of over 2000 people, hospitalization of over 50 000 and significant exposure of over 320 000 people in Bhopal, India. The release of the gas occurred early on the morning of 3 December 1984, lasted only 40 min and over a period of not more than 2 h blanketed an area of approximately 60 km^2. The gas covered a significant portion of the city of Bhopal, which had a population of nearly a million people (McConnell *et al.*, 1987). The disaster in Bhopal was apparently caused by the introduction of water into an MIC storage tank resulting in a severe exothermic reaction. Several safety precautions were reputed to have failed and some 18 000 kg of MIC and possibly toxic reaction products escaped into the atmosphere (Lepkowski, 1985).

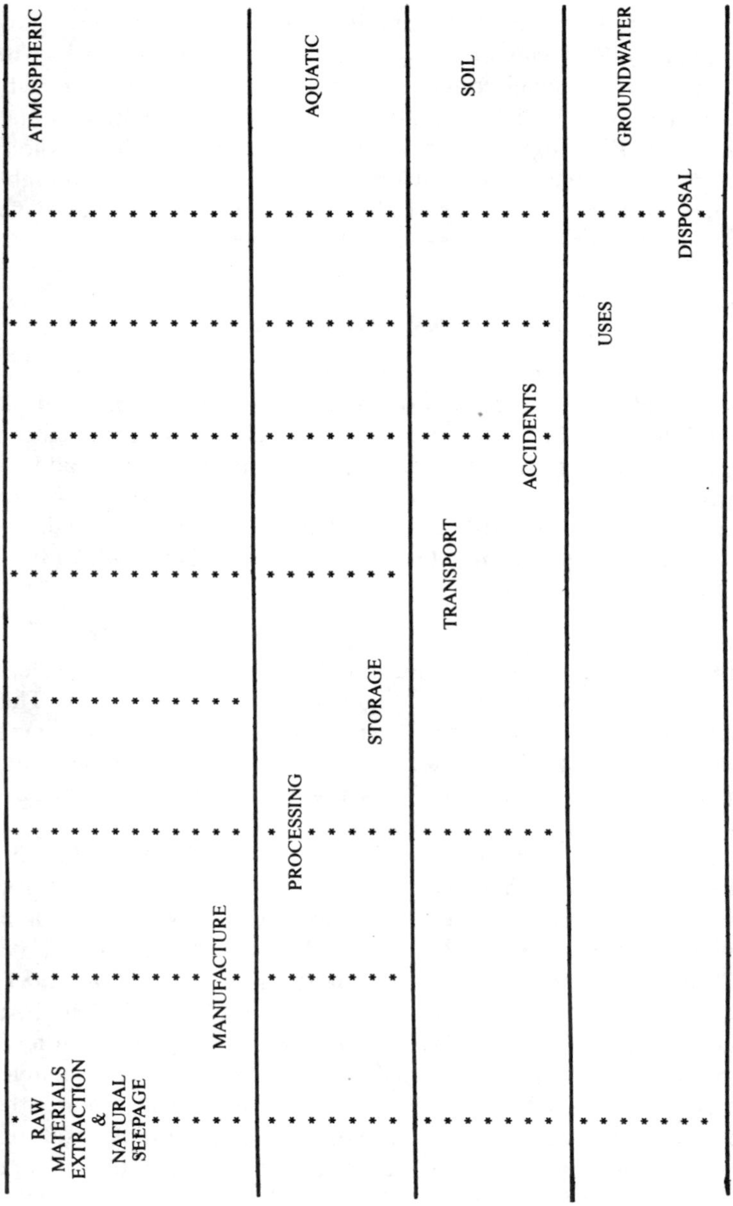

ATMOSPHERIC

AQUATIC

SOIL

GROUNDWATER

RAW MATERIALS EXTRACTION & NATURAL SEEPAGE

MANUFACTURE

PROCESSING

STORAGE

TRANSPORT

ACCIDENTS

USES

DISPOSAL

Fig. 3.2. Human activities leading to the release of chemicals into the environment. Adapted from Brown and Bomberger (1983).

MIC is used in the manufacture of pesticides particularly aldicarb, carbaryl, carbofuran and methomyl. It is flammable and highly reactive with water. United States production has been estimated at between 30 to 35 million pounds per year (Anon, 1984). The initial symptoms experienced by the exposed individuals in Bhopal were those of acute fulminating inflammation of the respiratory tract, with death ascribed to severe pulmonary oedema and haemorrhage. Implicated in the background to the disaster were faulty, badly maintained equipment, imperfectly trained personnel, poor attention to previous safety violations, and the proximity of the Union Carbide plant to commercial and residential premises (Davidar, 1985).

Disregard and deliberate contravention of safety rules concerning the operation of the number four reactor at the Chernobyl nuclear power station located 90 km north of Kiev in the Ukraine resulted in a major uncontrolled release of radionuclides into the atmosphere and thence throughout the northern hemisphere (UNSCEAR, 1989). Significant radionuclide releases due to reactor accidents have also occurred at Windscale (now Sellafield) in Cumbria, UK, in 1957 and at Three Mile Island, USA, in 1979 (UNSCEAR, 1982).

In November 1986, a fire at a Sandoz Ltd storehouse at Schweizerhalle, an industrial area near Basel, Switzerland, resulted in chemicals being discharged into the River Rhine causing massive kills of benthic organisms and fish (Rich, 1986; Deininger, 1987). The majority of the more than 1300 tonnes of stored chemicals were destroyed in the fire, but large quantities were introduced into the atmosphere, into the Rhine through run-off of the fire-fighting water and into the soil and groundwater at the site. The principal chemicals stored were organophosphorus insecticides such as disulfoton, etrimfos, fenitrothion, parathion, propetamphos and thiometan, mercury and zinc based pesticides, and other pesticides including DNOC, oxadixyl and metoxuron (Capel *et al.*, 1988). Fortunately, this accident has not proved to be the long-term ecological disaster originally feared.

Hazardous waste

American academic institutions collectively generate less than 1% of the USA's hazardous waste compared with 6% from the paper manufacturing industry, 8% from metals production, 10% from machinery, 59% from the chemical industry and a residual 16% from other sources (EPA, 1979). Even so, totals at individual institutions can be surprisingly large. For instance, the University of Illinois at

Urbana-Champaign disposed of 27 500 kg of waste chemicals in 1984. This was made up of 2104 chemicals and chemical mixtures, collected in 7300 containers (Ashbrook & Reinhardt, 1985).

In spite of this apparently diverse collection, general patterns in the nature of many academic institutions' wastes could be seen, with spent organic solvents representing the greatest proportion of hazardous wastes. Other major contributors were spent acids and bases and unwanted stock chemicals. A small proportion of laboratory wastes were principally comprised of spent toxic metals, degraded stock chemicals, contaminated laboratory apparatus, chemicals that react with air or water, potentially explosive chemicals, cyanides and insecticides, pesticides, polychlorinated biphenyls and small, used gas cylinders. Perhaps one of the major problems faced in dealing with hazardous academic wastes is the likely change in composition and toxicity with each new research project and experiment.

One of the most severe and well publicized cases of environmental contamination associated with hazardous waste disposal is that of Love Canal, Niagara Falls, New York. Officials of the Hooker Chemicals and Plastics Corporation have acknowledged that from 1942 to 1953 the company disposed of some 21 800 tons of chemical wastes in a trench in what was then a sparsely populated portion of the southeast corner of the city of Niagara Falls. The area contained the remnants of a canal that was to supply power to a utopian industrial community called Model City, which was to have been built in the early 1890s by entrepreneur William T. Love.

In 1955, the Niagara Falls Board of Education, which had acquired the site, opened an elementary school on a plot adjacent to the waste burial site. By 1966, residential development had eliminated all surface evidence of the earlier excavation. From the autumn of 1975 to the spring of 1976, heavy precipitation resulted in an unusually high local ground water condition. This caused subsidence, creating ponds of chemically contaminated surface water, and transporting chemical waste to nearby residences. On 7 August 1978, President Carter issued an executive order declaring that a man-made state of emergency existed at Love Canal. By 1987, approximately 100 million $US had been spent on remedial treatment, relocation of residents and environmental and human health investigations (EPA, 1982; Deegan, 1987*a,b*).

Among the lessons learned was that the environmental assessment of a site should be performed on the basis of integrated, sequential studies. It should include a comprehensive hydrogeological investigation of the site to determine the potential for contaminant migration and the likely direction of contaminant movement. Multimedia environmental monitoring studies should also be a part of an assessment to identify site and specific contaminants and to determine the likely extent of contaminant movement from the site. Field sampling should ensure that the gradients of all contaminant plumes, in all media, are identified and traced directly from the source to their ambient levels or boundaries of non-detection.

After discoveries of waste dump sites such as at Love Canal and the so-called Valley of the Drums in Bullitt County, Kentucky, a federal 'Superfund' was established in December 1980 to pay for cleaning up abandoned, uncontrolled hazardous waste sites. From 1981 to 1985 Superfund was financed to the extent of nearly 1·6 billion US$ and supplemented by 10% matching grants from the individual states. Up to 86% of the federal money was raised from special taxes imposed on petroleum, petrochemical and chemical manufacturing companies. Before the expiry of Superfund in September 1985, the EPA identified 20 766 potentially hazardous waste sites in the USA. These included private and government-owned chemical, mining and radioactive waste sites as well as municipal dumps containing hazardous materials (Josephson, 1985). Table 3.1 lists some contaminants found at Superfund sites.

Table 3.1
Some contaminants found at Superfund sites

Chemical	*Average percentage occurrence*	*Average concentration (ppb)*
Lead	51·4	$\simeq 3{\cdot}09 \times 10^5$
Cadmium	44·7	2 185
Toluene	44·1	$\simeq 1{\cdot}12 \times 10^6$
Mercury	29·6	1 379
Benzene	28·5	16 582
Trichloroethylene	27·9	$\simeq 1{\cdot}03 \times 10^5$
Ethylbenzene	26·9	$\simeq 5{\cdot}4 \times 10^5$
Benzo[*a*]anthracene	12·3	$\simeq 1{\cdot}48 \times 10^5$
Bromodichloromethane	7·0	20
Polychlorinated biphenyls	3.9	$\simeq 1{\cdot}28 \times 10^5$
Toxaphene	0·6	12 360

Data from Josephson (1985) (after Eckel *et al.* (1985)).

Sites were rated according to their potential for releasing hazardous wastes that damage human health and the environment. The National Priorities List (NPL) used a mathematical rating scheme known as the Hazard Ranking System (HRS) (Federal Register, 1982). Sites were also rated according to the potential for migration of wastes to groundwater, leachate toxicity, and air and soil contamination. The degree of hazard of the waste material was also taken into account.

If points awarded under the HRS on the basis of such characteristics exceeded a total of 28·5, then the site was proposed for inclusion on the NPL. Further investigation and public comment were required before final inclusion.

A detailed evaluation and explanation of EPA's HRS has been provided by Wu and Hilger (1984) and the EPA's risk assessment ethos has been outlined by Yosie (1987). A special issue of the international journal *The Science of the Total Environment,* **51** (May 1986) has addressed the topic of risk management in chemical safety. Research focusing on humans as receptors of pollutants has been reviewed by Ott (1985).

The EPA also proposed a classification of hazardous industrial wastes partly based on generic characteristics such as ignitability, corrosiveness, reactivity and toxicity (Table 3.2). The control of hazardous chemical spills in the UK has been reviewed by Cumberland (1982). It was stressed that good communications are a basic requirement for any emergency response capability. The marking of packages and vehicles with an appropriate telephone number, a statutory requirement for certain products, was a fundamental principle in the Chemical Industry Scheme for Assistance in Freight Emergencies (CHEMSAFE). Good cooperation between industry and local and central government was believed to be a key factor in both reducing the number of hazardous chemical spills and minimizing their environmental impact, should they occur. The World Health Organization has also sponsored a technical manual for the safe disposal of hazardous wastes with special emphasis on the problems and needs of developing countries (WHO, 1987). The seventh edition of the standard reference work on industrial chemical hazards has been published (Sax & Lewis, 1988).

The deactivation of hazardous chemical wastes has been considered by Tucker and Carson (1985). Deactivation by chemical conversion results in the waste, or a constituent, being transformed to at least one substance that is less hazardous than, and chemically different from,

Table 3.2
Classification of hazardous waste as proposed by the US Environmental Protection Agency

Generic characteristics

A. *Ignitable*
   1. Flash point <60°C (<140°F)—Pensky-Martens closed Cup Tester ASTM Std D-93-79 or Setaflash Closed Tester ASTM Std D-3278-78
   2. Solids with high ignitable potential and when ignited burn so vigorously and persistently that it creates a hazard
   3. Ignitable compressed gas (49 CFR 173.300)
   4. Oxidizer (49 CFR 173.151)

B. *Corrosive*
   1. pH $\leq$ 2·0—pH meter—*Methods for Analysis of Water and Wastes* (EPS 600/4-79-020, March 1979)
   2. pH $\geq$ 12·5
   3. Corrosion rate 0·25 in/yr (6·35 mm year^{-1}) at 130°F (54·4°C) NACE Std TM-01-69

C. *Reactive*
   1. Unstable (autopolymerize, thermal, shock, strong oxidizers) (shock instability test as required for Class A explosives)
   2. Reacts violently with water
   3. Forms potentially explosive mixture with water
   4. Generates toxic fumes when mixed with water in a quantity sufficient to present a danger to human health or the environment
   5. Cyanide or sulphide bearing
   6. Capable of detonation
   7. Forbidden, Class A (49 CFR 173.53) or Class B (49 CFR 173.88) explosive

D. *Toxic*
   1. Based on liquid phase or extract from Toxicant Extraction Procedure (TEP) as presented
   2. Concentration 100 × $\geq$ Primary Drinking Water Standard

E. A Generator is any person who produces more than 1000 kg in any one month with intent to decrease to 100 kg month^{-1} over a 2 to 5 year period. 1 kg month^{-1} limit for 'acutely hazardous chemical'

Data from Englande and Reimers (1980).

the original material. Usually, methods for removing hazardous constituents result in a hazardous concentrate, and a reduced total volume of waste. Waste may be stabilized by fixing in cement or other material to form a hardened material, or may be encapsulated in an inert substance. Some chemical conversion processes for hazardous wastes are listed in Table 3.3. Processes for the removal of hazardous

Table 3.3
Chemical conversion process for hazardous wastes

Process	*Waste type*
Wet oxidation	Variety of organic compounds
Ozonation	Phenols, cyanides, and organic lead compounds in wastewaters, compounds in air
Molten-salt combustion	Variety of organic compounds
Electrochemical oxidation	Cyanate, thiocyanate, acetate, phenols, cresols
Treatment with formaldehyde	Chromium(VI), cyanide ion, metal–cyanide complexes
Catalytic reduction with metal powder	Chlorinated organic compounds in wastewaters
Catalytic hydrogenation	Polychlorinated hydrocarbons
Dechlorination	Polychlorinated organic compounds
Destruction by microwave plasma	Pesticides, polychlorinated biphenyls
Hydrolysis	Organophosphorus pesticides, carbamate pesticides
Neutralization	Strong acids, alkalis

Data from Tucker and Carson (1985).

Table 3.4
Processes for the removal of hazardous constituents

Process	*Hazardous constituents*
Adsorption on carbon	Wide variety of compounds in wastewaters and air
Ion exchange	Ions in wastewaters
Cementation	Metal ions in wastewaters
Precipitation	Heavy-metal cations, certain anions in wastewaters
Liquid–liquid extraction	Organic compounds, metal ions
Ultrafiltration	Colloidal particles, large molecules in liquids
Reverse osmosis	Most solute molecules, ions in wastewaters
Electrolytic reduction	Metal ions

Data from Tucker and Carson (1985).

Table 3.5
Methods and applications of stabilization

Stabilization method	*Application*
Solidification with cement	Sludges, contaminated soil, various metal salts, low-level radioactive waste
Solidification by lime-based processes	Flue gas desulphurization wastes, other inorganic wastes
Solidification with thermoplastic materials	Radioactive wastes
Solidification with organic polymers	Sludges, radioactive sludges
Encapsulation	Sludges, liquids, particulate matter
Solidification by self-cementation	Flue gas desulphurization cementation wastes, other large proportions of calcium sulphate or calcium sulphite
Vitrification	Extremely hazardous wastes, radioactive wastes

Data from Tucker and Carson (1985).

constituents and methods and applications of stabilization are summarized in Tables 3.4 and 3.5, respectively. It was argued that deactivation offers many advantages to the industrial community including lowered exposure to workers, a decreased need for storage capacity, lower handling and transport costs and a smaller impact on the environment. Remedial action technologies have been addressed in a five-part series of articles commencing with one by Chudyk (1989).

3.3 MANUFACTURING PROCESSES

Examination of manufacturing processes often reveals a pattern of exposure pathways for the physical elimination of potential pollutants, and also a pattern of chemical processes and transformations that can lead to both specific and generalized expectations about the nature of emissions. Taking a schematic view of a generalized industrial plant, as is illustrated by Fig. 3.3, it is possible to anticipate the likelihood of fugitive air emissions, water pollution and landfill material leachates.

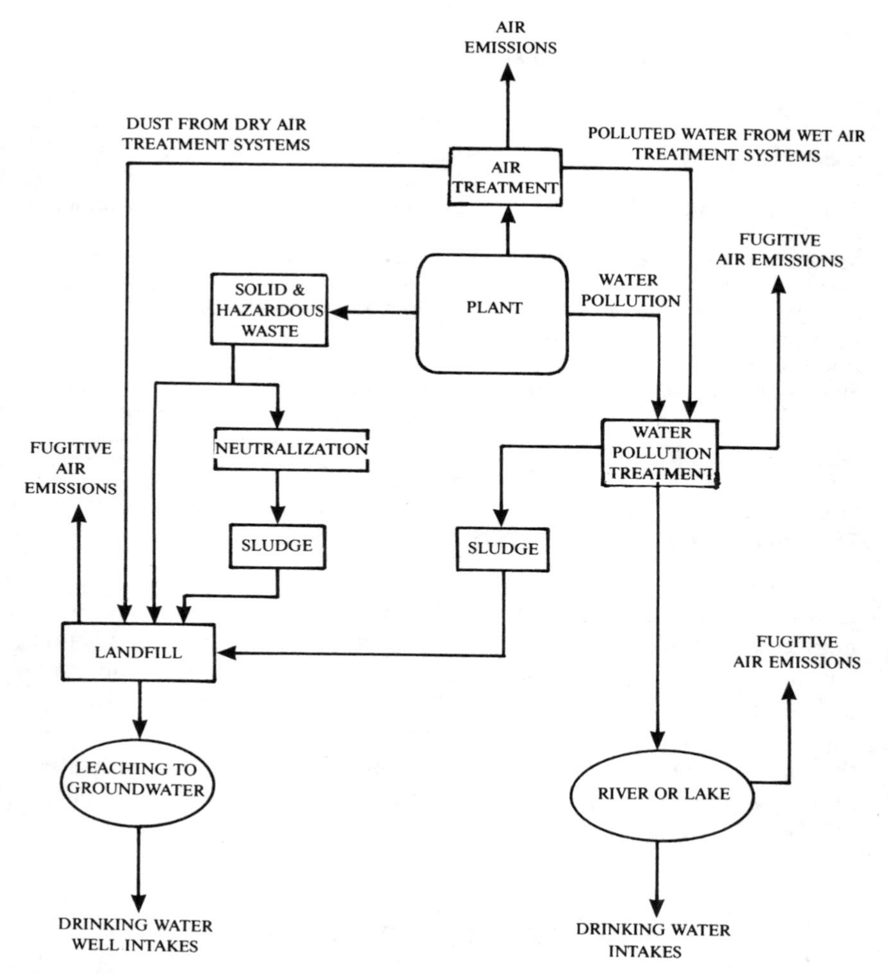

Fig. 3.3. Pollutant flows and exposure pathways for a generalized industrial plant. Reproduced from Barry (1987).

Generic character

If in turn one looks at the chemistry of reactions occurring in the manufacturing process it is apparent that all chemical products are manufactured from some precursor by definite processes that dictate the chemistry of the transformation and which are designed to optimize the formation of the required substance. The chemistry and conditions of a process give it a generic character. Thus, when a destined generic process is applied to specified precursors, an opportunity exists for the prediction of the nature of the product.

Despite the apparent diversity of chemical products, many organic chemicals are derived from a few primary precursors via a limited number of synthetic routes that employ processes with generic character (Wise & Fahrenthold, 1981; Herrick & King, 1979).

An econometric model of industrial products and processes showed that the principal organic chemicals used are those involved in the manufacture of the monomeric precursors of the commercially important plastics and synthetic fibres (Wise & Fahrenthold, 1981). It was observed that these chemicals derive from only eight precursors, namely benzene, toluene, *o*-xylene, *p*-xylene, ethylene, propylene, butane/butene and methane. These precursors in turn were predominantly derived from crude oil fractions, natural gas and coal tar distillates.

EPA designated 'priority pollutants' were limited to 129 in number and in 1981 included 116 organic chemicals of which 88 were named specifically. A priority pollutant profile of the organic chemical industries was given by Wise and Fahrenthold (1981). The latter authors argued that by superimposing the synthetic routes to priority pollutants over the synthetic routes employed by industry in the manufacture of organic chemicals, it is possible to identify products of coincidence. Product processes that generate priority pollutants were therefore deemed identifiable.

Such an approach simplifies consideration of the organic chemical industry by focusing attention on the manufacturing processes of a limited number of products that are most likely to generate priority pollutants. Critical precursor–generic process combinations are inherent in such production processes. Important precursor–generic processes that generate priority pollutants are indicated in Table 3.6.

Wood pulp processing

Consideration of the composition and sources of pulp bleaching wastes serves as one example of how it is possible to anticipate the nature of contaminants likely to be released to the environment from a particular manufacturing process.

Wood is the essential raw material for the production of chemical pulp, being composed of cellulose, hemicelluloses, lignin and extrac-

Table 3.6
Critical precursor–generic process combinations that generate priority pollutants

Precursor	*Oxidation products*	*Chlorination products*	*Generic process (nitration products)*	*Diazotization products*	*Reduction products*
Benzene	Phenol	Chloroaromatics Chlorophenols	Nitroaromatics Nitrophenols		
Toluene	*o,m*-Cresol		Nitroaromatics		
Xylene	2,4-Dimethylphenol		2,4-Dimethylphenol		
Naphthalene		2-Chloronaphthalene			
Phenol		Chlorophenols	Nitrophenols		
Cresols		4-Chloro-*m*-cresol	4,6-Dinitro-*o*-cresol		
Chloroanilines				Chlorophenols Chloroaromatics Aromatics	
Nitroanilines				Nitrophenols Nitroaromatics Aromatics	
Nitrobenzene				*N*-Nitrosodiphenylamine[a]	Aniline (diphenylamine[a])
m-Chloronitrobenzene				Benzidines[b]	1,2-Diphenylhydrazines[b]
Ethylene		Chlorinated C 2's Chlorinated C 4 Chloroaromatics			
Propylene	Acrolein	Chlorinated C 3's			
Methane		Chlorinated methanes			

[a] Derived directly from aniline, or indirectly via phenylhydrazine, diphenylamine is one of three secondary amines that are precursors for nitrosamines, when exposed to nitrites (as in diazotization) or NO_x.
[b] Diphenylhydrazines rearrange to benzidines under acid conditions (as in diazotization).
Data from Wise and Fahrenthold (1981).

tives (Sjöström, 1981). Cellulose is a *linear* polysaccharide made up of β-D-glucopyranose units linked by (1–4)-glucosidic bonds. About 40% of most wood is cellulose of molecular weight in excess of 10 000. Wood hemicelluloses are *branched* different carbohydrate units of lower molecular weight. In softwoods, galactoglucomannans (15–20%), arabinoglucuronoxylan (5–10%) and arabinogalactan (2–3%) are the most common types of hemicellulose. In hardwoods, glucuronoxylan (20–30%) and glucomannan (1–5%) are the predominant variety.

Lignin is an aromatic polymer of potentially infinite molecular weight. Softwood lignin is essentially a branded polymerization product of coniferyl alcohol with an aromatic content, expressed as monomeric phenol, of around 51%. Hardwood lignin is formed by copolymerization of coniferyl and sinapyl alcohols with a relative ratio varying between 4:1 to 1:2 (Kringstad & Lindström, 1984).

Those components of wood, which can be removed by organic solvents (1·5–5%) are termed 'extractives'. These may be classified as aliphatic extractives (fats and waxes); terpenoid compounds (only in softwoods) and resin acids (e.g. pimaric acid, abietic acid and dehydrabietic acid); and thirdly, phenolic extractives such as hydrolysable tannins, flavanoids, lignans, stilbenes and tropolines.

Most chemical wood pulping is carried out either by the kraft (sulphate) or by the sulphite process, the purpose of which is to remove lignin in order to facilitate fibre separation and papermaking properties (Rydholm, 1965). Of the two processes, the kraft system is used predominantly. This entails treating wood chips at 160–180°C with a liquor containing sodium hydroxide and sodium sulphide in order to cleave ether bonds in the lignin. Lignin degradation products are then dissolved in the alkaline pulping liquor, resulting in some 90–95% of the lignin being removed from wood at this stage. Aldehydic end groups of cellulose and hemicelluloses are sensitive to alkali and also degrade (the peeling reaction), resulting in a number of organic acids being dissolved in the pulping liquor. The peeling reaction ceases when the end group rearranges to an alkali stable *m*-saccharinic end group. Wood extractives are also dissolved or dispersed in the kraft pulping liquor. With softwoods, most extractives are recovered as sulphate turpentine (lower terpenes) and tall oil (fatty and resin acids).

In the kraft process, more than 55% of the total weight of wood is dissolved in the pulping liquor. After separation from the pulp, the spent liquor is evaporated to a high concentration and then burned to recover energy and inorganic chemicals (Kringstad & Lindström, 1984). Figure 3.4 illustrates the kraft process. In contrast, the sulphite process contains sulphur, magnesium or calcium, and lignin is solubilized by sulphuration at elevated temperatures.

Figure 3.4 also illustrates the principles of bleaching, required to remove the residual 5–10% of the original lignin, which cannot be pulped further without severe degradation of the polysaccharide fraction. Most material is dissolved from the pulp during the C (elemental chlorine) and E (alkali extraction) stages. In spent chlorination liquor about 70% of the organically bound chlorine is present as high-relative-molecular-mass material ($M_r > 1000$) compared with 95% in alkali extraction liquor (Lindström *et al.*, 1981). The elemental composition of high M_r material has been represented as $C_9H_9O_4Cl$, $C_{10}H_{14}O_7Cl$ and $C_9H_{10}O_8Cl$ for spent chlorination liquor, and typically, $C_{14}H_{15}O_8Cl$ and $C_{14}H_{10}O_9Cl$ for alkali extraction liquor (Bennett *et al.*, 1971). Although such materials are probably biologically inactive because they cannot penetrate living cell membranes, their environmental impact may be considerable. This is because they contain chromophoric structures that cause bleaching plant effluents to increase light absorption in receiving waters. In addition, low M_r biological and degradation products of high M_r substances may have biologically toxic properties.

Of the organically bound chlorine in spent chlorination liquor, some 30% is of low relative molecular mass ($M_r < 1000$). Alkali extraction liquor contains approximately 5% low M_r. Compounds include oxalic acid, trichloropropenoic acid, chloro-2-thiophenic acid, 2,4-dichlorophenol, tetrachloroguaiacol, dichlorovanillin, chloroform, trichloroethene and pentachlorobenzene. Phenolic compounds most frequently identified in spent liquors from softwood kraft pulps are chlorinated phenols, chlorinated guaiacols, chlorinated vanillins and chlorinated catechols. Additionally, in spent liquors from birch kraft pulps are chlorinated syringols and chlorinated syringaldehydes (Voss *et al.*, 1981).

Spent chlorination and alkali extraction liquors from the bleaching of softwood kraft pulp are mildly toxic to fish and other aquatic organisms (Walden, 1976; Walden & Howard, 1977). Most toxic is 3,4,5-trichloroguaiacol in alkali extraction liquor, especially in combination with tetrachloroguaiacol and mono- and di-chlorodehydroabietic acid. Although normally of lower absolute

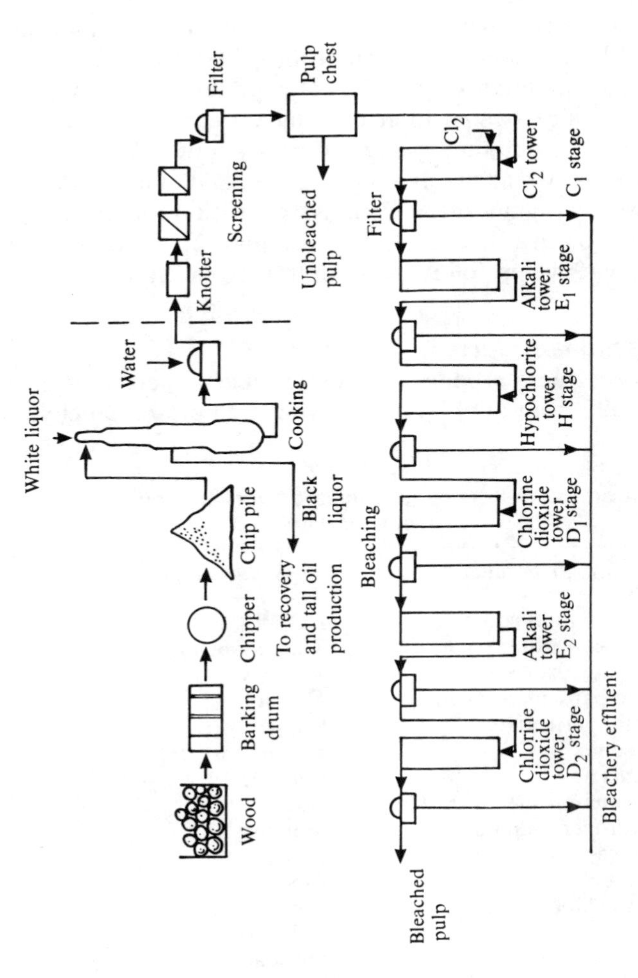

Fig. 3.4. Kraft pulp mill and conventional bleaching plant. Reproduced from Kringstad and Lindström (1984) (after Rydholm (1965)).

toxicity, the greater volumes of chlorination liquor result in a greater total toxic effect (Sameshima *et al.*, 1979). Many compounds released are mutagenic in their pure state. Of particular significance are chlorinated derivatives (chloroacetones) of 3-chloro-4-dichloromethyl-5-hydroxy-2(5*H*)-furanone, and 2-chloropropenal.

Thirty-five air pollutants and 143 water pollutants, including those derived from sludge composition data have been identified as being associated with the pulp and paper industry (Barry, 1987). Through a screening process, which involves aggregation and elimination of those pollutants with insignificant emissions, or with little or no known environmental or health effects, this list may be reduced to 12 air pollutants and 20 water pollutants as presented in Table 3.7.

Coal liquefaction products

Crude synthetic fuels produced by direct coal liquefaction generally differ chemically from petroleum in having greater aromaticity, higher

Table 3.7
Important pollutants associated with the pulp and paper industry

Air pollutants	*Water pollutants*
Carbon monoxide	BOD5
Nitrogen oxides	Total suspended solids
Sulphur oxides	Arsenic
Fine particulates <10 μm	Cadmium
Total particulates	Copper
Arsenic	Chromium (III)
Benzene	Cyanide
Chromium (hexavalent)	Lead
Hydrogen sulphide	Mercury
Mercaptans	Nickel
Nickel	Zinc
Tetrachloroethene	Chloroform
	Chlorine
	Fatty acids
	Phenol
	PCBs
	Resin acids
	Tetrachloroethylene
	2,4,6-TCP

Data from Barry (1987).

concentrations of polar polycyclic aromatic hydrocarbons (PAHs), phenols and nitrogen containing compounds, and shorter and more numerous alkyl side chain substituents and aromatic rings (Crynes, 1981).

When coal-derived oil is released into water, the oil constituents immediately begin to partition according to their chemical and physical properties. Volatile components quickly evaporate from a floating slick, while soluble components dissolve in the water column. Insoluble components emulsify, sink directly or sorb to particles and eventually accumulate in bottom sediments. The water-soluble fraction (WSF) can contaminate drinking water supplies and be toxic to aquatic life. However, normally, most WSF compounds in coal liquids are readily degraded by bacteria.

Giddings *et al.* (1985) used outdoor and aquaria microcosms to investigate the pattern of contamination that followed a spill of coal-derived crude oil containing 3% phenolics. Dissolved oil concentrations reached a maximum within 24 h and thereafter declined. Unsubstituted phenol and cresols were degraded most rapidly, and highly alkylated phenols were more persistent. The maximum concentration of dissolved oil expected after a spill in standing water could be estimated from the reactive volumes of oil and water and the WSF of the oil. For example, a spill of $300\,m^3$ (80 000 gallons) of oil containing 3% phenolics into a typical embayment $1\,km^2$ in area and 1 m deep ($10^6\,m^3$) would result in a theoretical maximum concentration of $(300 \times 0{\cdot}03)/(10^6) = 9\,\mu g\,ml^{-1}$ dissolved phenolics. Figure 3.5 illustrates the pattern of dissolution and evaporation of coal liquids in water. Phenols and anilines dissolve more rapidly and evaporate more slowly than do light aromatics. When petroleum is spilled, evaporation removes much of the soluble material before it dissolves in the underlying water (Samiullah, 1985). In contrast, most of the soluble components of a spilled coal liquid dissolve faster than they evaporate, resulting in relatively high concentrations of dissolved and potentially toxic dissolved oil.

Organic compounds in wastewater

A series of reviews has given a complete picture of the effort to characterize waters of all types in the 1980s (De Walle *et al.*, 1981, 1982; Adams *et al.*, 1983–1985). More recently, some 4000 wastewater samples from 40 industrial categories and publicly owned treatment works (POTWs) were analysed for 114 EPA organic priority pollu-

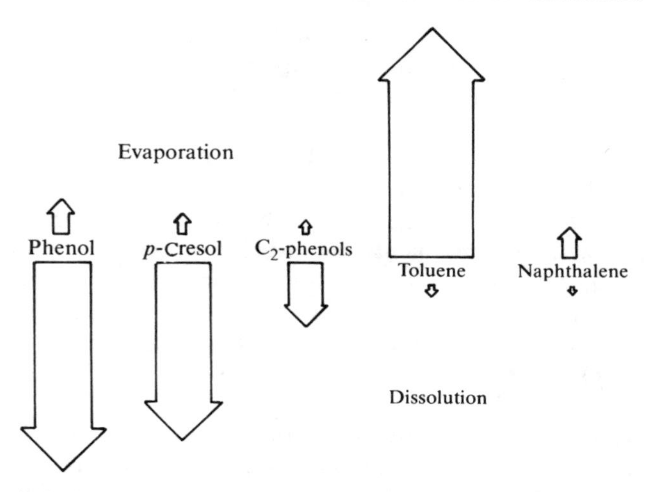

Fig. 3.5. Relative mass fluxes from a coal-derived liquid layer in water. Mass flux is proportional to width of arrow; extent of evaporation and dissolution is proportional to length of arrow; coal liquid layer is 1·4 mm thick; wind speed 2 cm above the oil (coal liquids) is 0·2 m/s. Reproduced from Giddings *et al.* (1985).

tants, using the technique of gas chromatography/mass spectrometry (Shackelford & Cline, 1986). Among the 50 most frequently occurring compounds overall, only 14 were priority pollutants. For the POTWs only, 16 priority pollutants were in the top 50, with the distribution of the top 14 essentially the same as in the overall data. For petroleum refining and iron and steel manufacture, a different distribution of priority pollutants was noted with respect to the other compounds. In any industrial category, the range of chemicals found was smaller than the total range of chemicals across all the categories studied.

It was observed that although there are some priority pollutants that are important for each industry, they make up only 25% of the 50 most frequently occurring compounds. In fact, 1% of the total number of compounds identified accounted for more than 32% of the total number of compound occurrences. However, of the total number of discrete compounds identified, 40% accounted for only 2% of the total number of compounds occurring. Figure 3.6 shows some of the largest compound classes, in terms of frequency of occurrence and numbers of class members identified. Clearly, there are whole classes of analysed compounds containing no priority pollutants whatsoever. Table 3.8

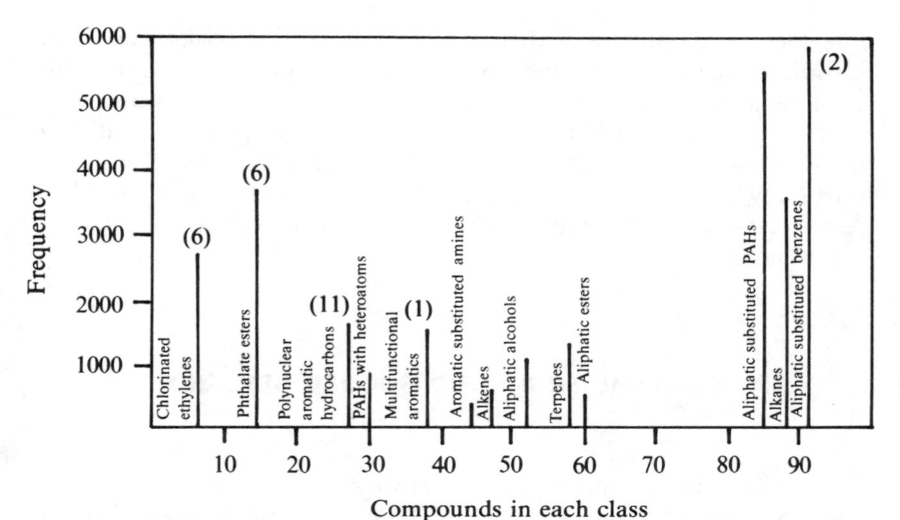

Fig. 3.6. Distribution of organic compound classes in industrial and publicly owned treatment works wastewaters. The number of priority pollutants in each class is given in parentheses. Reproduced from Shackelford and Cline (1986).

Table 3.8
The 10 largest peaks in GC/MS analyses of 4000 wastewater samples from 40 industrial categories in publicly owned treatment-works wastewaters

Industrial category	*Compound*	*Frequency of occurrence*	
		Overall	*Specific industry*
Soaps and detergents	Phenyltridecane	5	5
Soaps and detergents	Undecanone	9	7
Gum and wood	3,4-Dimethyl-2,5-furanedione	7	5
Gum and wood	*O,O*-Diethyl-*S*-ethylphosphorodithioate	5	4
Oil and gas extraction	Benzenebutanoic acid	7	5
Oil and gas extraction	Ethyl phenylacetate	13	10
Oil and gas extraction	Benzyl chloride	10	7
Rum production	2,6-Dimethylpyrazine	5	5
Rum production	3-Indoleacetamide	5	4
Rum production	Acetylpyrrole	5	4

Data from Shackelford and Cline (1986).

lists the 10 compounds with the highest analytical peaks that were recorded, along with the industries with which they were associated. Looking at the first row in this table, it can be seen that phenyltridecane was found in industrial wastewater five times in all, but all five occasions were related to the soap and detergent industry. Therefore, this knowledge can enable researchers to better target available resources when considering likely contaminants from a given manufacturing process.

3.4 PHYSICAL FORM AND SOURCE DESCRIPTION

Chemicals may be presented to environmental compartments in gaseous, solid or liquid form. Table 3.9 cites common physical forms of substance encountered by the compartmental phases land, air, ground water and surface water (rivers, lakes and oceans).

Knowing the physical form, it is possible to anticipate the behaviour and persistence of categories of chemicals.

Environmental fate models require information on the temporal and spatial distribution of chemical releases. Sources can be described in terms of their dimensionality and releases in terms of their temporal distribution. Releases may be continuous and approximately uniform (e.g. municipal landfill), intermittent but predictable (e.g. industrial emissions), cyclic or random.

Table 3.9
Forms of substances in receiving media

Media	*Solids*	*Liquids*	*Gases*	*Combinations*
Air	Particulate	Vapour particulate	Gas	Adsorbed gas or liquid
Surface water	Suspended	Dissolved	Dissolved	Co-solution
Groundwater	Dissolved	Dissolved	Dissolved	
Land	Particulate	Contained	Contained	
	Bulk	Absorbed		
	Contained	Adsorbed		

Data from Brown and Bomberger (1983).

Categories of sources and releases are outlined as follows:

Dimensionality

Point sources	(smoke stacks).
Line sources	(the aggregated emissions from vehicles moving along a road).
Area sources	(conglomerate domestic fuel emissions).
Volume source	(nebulous concept—perhaps an example could be photochemical smog).

Temporal distribution

Accidents	(instantaneous: kg per event).
Point sources	($kg\,s^{-1}$).
Line sources	($kg\,s^{-1}\,m^{2}$).

3.5 ESTIMATING RELEASES

The *severity* (S) of a substance in a discharge from an industrial source is defined as the ratio of substance concentration, either at the source or at some ambient point of interest, to a maximum specified 'safe' concentration level (Leadbetter & Tucker, 1981). A source is considered 'clean' unless S is expected to exceed unity in more than a given and acceptably small proportion of time, otherwise it is classed as 'dirty'. Classification is based on such parameters as stack emission characteristics and meteorology. Releases may be estimated by a variety of methods, principally:

1. Measurement—either by direct sampling or calculation from known application rates.
2. Materials balance—can reduce to the simple mass balance where the measured mass of a chemical in products leaving the plant is subtracted from the raw material entering the plant to yield the loss.
3. Mathematical models—usually relatively simple.
4. *Ad hoc*—using the most obvious suppositions and calculations for a given situation. These include rule-of-thumb bounds on lost fractions of production likely to be economically acceptable.

3.6 ENVIRONMENTAL PATHWAYS

Derived from a study sponsored by the Umweltbundesamt in Berlin (Frische *et al.* 1979*a,b*), Fig. 3.7 was proposed as a schematic representation of the most important pathways for an organic chemical in the environment by Klopffer *et al.* (1982). In this diagram, the environmental media have been subdivided into sectors with likely molecular exchanges arrowed. For example, an exchange will probably take place between A_1 (gaseous distribution in the troposphere) and A_3 (adsorbed as aerosol), or between B_3 (hydrosol) and C_2 (sediment). For highly persistent compounds (no efficient sinks in any sector except A_2 and B_5), Fig. 3.7 may be used as the basis of an environmental model for predicting approximate concentrations (C_x) in different phases by solving equations of the type:

$$\frac{dC_x}{dt} = I_x + \sum_i k_i C_i - C_x \sum_j k_j \ (=0 \text{ in steady state})$$

where I_x is the input rate into sector x from the '*technosphere*' (per volume), k_i are rate constants of environmental transfer processes into sector x (first-order processes assumed), C_i is the concentration of the substance in the sector donating the substance to x, and k_j are rate

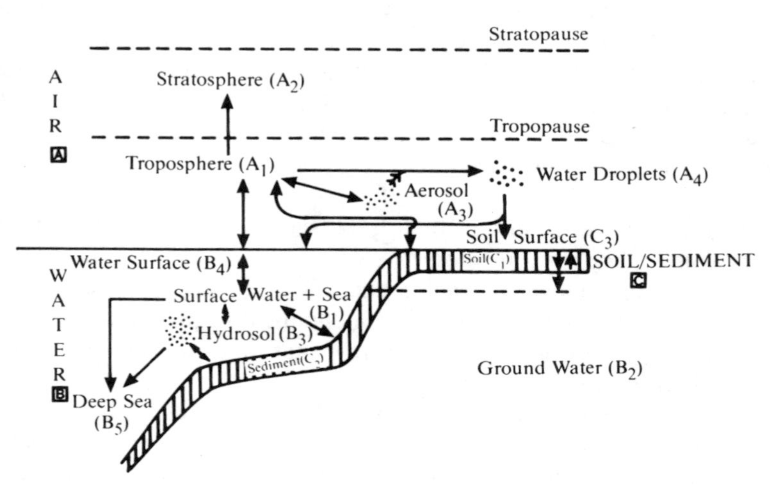

Fig. 3.7. Important pathways for organic chemicals in the environment. Reproduced from Klopffer *et al.* (1982).

Table 3.10
Major pathways of chemicals

	Pathways between sectors[a]	*Mechanism*	*pc or env.*[b]
	$A_1 \rightarrow A_2$	Turbulent diffusion	env.
	$A_1 \rightarrow A_4$	Dissolution gas → water	pc
	$A_4 \rightarrow A_1$	Volatility from aqueous solution	pc
A	$A_1 \rightarrow A_3$	Adsorption gas → aerosol	pc
	$A_3 \rightarrow A_1$	Volatility from adsorbed state	pc
	$A_3 \rightarrow A_4$	Condensation of water on aerosol	env.
A	$A_1 \rightarrow B_4$	'Dry deposition' (dissolution gas → water)	pc
	$A_1 \rightarrow C_3$	'Dry deposition' (adsorption)	pc
↓	$A_3 \rightarrow C_3 + B_4$	'Dry deposition' (of particles)	env.
B	$A_4 \rightarrow C_3 + B_4$	Rain	env.
+	and		
C	$A_3 \rightarrow A_4 \rightarrow C_3 + B_4$ (direct)	'Raining out'	
	$B_4 \rightleftharpoons B_1$	Turbulent diffusion and mixing	env.
	$B_1 \rightarrow B_5$	Turbulent diffusion	env.
B	$B_1 \rightarrow B_3$	Adsorption	pc
	$B_3 \rightarrow B_1$	Desorption	pc
	$B_3 \rightarrow B_5$	Sedimentation	env.
	$B_1 \rightarrow B_2$	Adsorption	pc
	$B_2 \rightarrow B_1$		
B	$B_4 \rightarrow A_1$	Volatility from aqueous solution	pc
↓	$B_4 \rightarrow A_3$	'White capping'	env.
A	$B_1 \rightarrow C_2$	Adsorption	pc
+			
C	$B_3 \rightarrow C_2$	Sedimentation	env.
C	$C_3 \rightarrow C_1$	Adsorption/desorption ('leaching')	pc
	$C_1 \rightarrow C_3$	Adsorption/desorption	pc
C	$C_1 \rightarrow B_2$	'Leaching'	pc
↓	$C_3 \rightarrow A_1$	Volatility from adsorbed state, aqueous solution (moist soil) and pure substance	pc
A			
+			
B			

[a] For notation of environmental media and sectors, see Fig. 3.7; reentry from stratosphere, deep sea, and some other quantitatively less important pathways have been neglected.
[b] pc: Determined by physicochemical properties of the substance and environmental limiting factors.
env.: Determined by environmental (e.g. meteorological) factors only; in some cases the assignment is still ambiguous.
Data from Klopffer *et al.* (1982)

constants of environmental processes by which the substance is transferred from *x* into some other sector.

The technosphere was distinguished from the natural environment primarily in that within this sphere not only were the mechanisms of distribution, degradation and accumulation well understood, but it was possible to control the concentration of a chemical once it had been released (Frische *et al.*, 1982). Possible transfer pathways in Fig. 3.7 are presented in Table 3.10.

The persistence of an organic substance has been defined by Frische *et al.* (1979*a,b*) as its half-life in the environment. In order to estimate this parameter, it is necessary to identify the most efficient mechanisms removing the compound irreversibly. These processes are called sinks and Table 3.11 lists 10 of the most efficient. Of these, Nos 1, 3 and 7 are deemed to have universal significance (Klopffer *et al.*, 1982). Others, such as No. 4, can only be effective if certain functional groups are present in the molecule. Five sinks can be quantified by physico-chemical measurement, whereas the others are defined predominantly by biochemical processes. The term sink is also more traditionally taken to mean that environmental compartment or phase in which a chemical ultimately accumulates.

Although sinks play a very important role in removing organic chemicals, additional information is needed in order to estimate their

Table 3.11
List of the most important sink processes for organic chemicals in the environment

No.	*Sink processes*[a]
1.	Photochemical oxidative degradation in the gas phase
2.	Photolysis
3.	Biotic degradation in water
4.	Hydrolysis
5.	Photochemical degradation in aqueous solution
6.	Biotic degradation/modification by water plants (e.g. algae)
7.	Biotic degradation in soil
8.	Photochemical oxidative degradation in the adsorbed state
9.	Degradation/modification by green terrestrial plants
10.	Anaerobic–biotic and reductive–abiotic degradation

[a] For organic chemicals, sink processes are defined as mechanisms by which the substance is irreversibly changed at the molecular level.
Data from Klopffer *et al.* (1982).

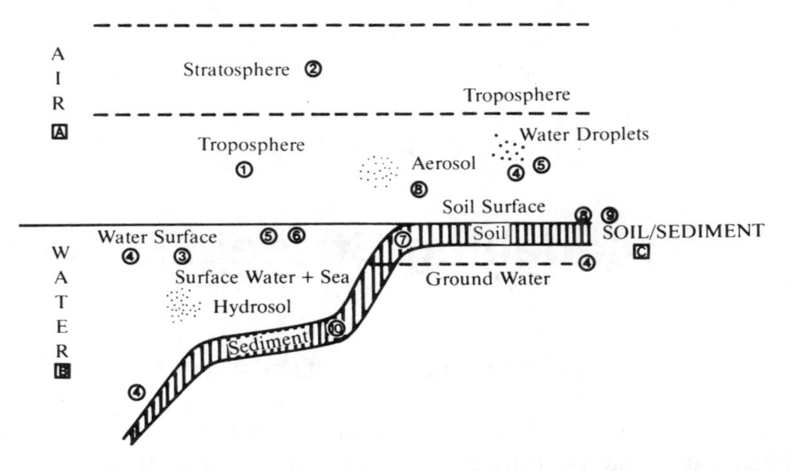

Fig. 3.8. Location of sinks in the environment. Reproduced from Klopffer *et al.* (1982).

half-life in the environment. This can be seen in Fig. 3.8, which shows the approximate location of the sinks listed in Table 3.11. Thus, photooxidation in aqueous solution (No. 5) can only occur near the surface of the water or in water droplets in the troposphere, for example. To be efficient, a sink has to be intrinsically reactive towards a specific chemical. Furthermore, the chemical has to reach the sink under environmental conditions. Although chemical reactivity can be calculated or determined in the laboratory, the ultimate sink is determined by the environmental mobility of the substance. Thus, calculation of the half-life of a chemical in the environment calls for knowledge of the rate constants of the individual sinks and the relevant phase transfer coefficients.

Bomberger *et al.* (1983) mentioned some simple screening techniques that could be used to quantify volatilization and leaching rates at the soil–air interface. These rates were then compared with estimates of transformation rate in order to determine a compound's overall fate and identify processes requiring further study. The movement of chemicals across the air, water and soil interfaces are further discussed by Neely (1980) and Chawla and Varma (1981/82), and in subsequent chapters dealing with the soil, aquatic and atmospheric environmental compartments.

Chapter 4

Modelling Strategies

4.1 MODELLING PHILOSOPHY

Neely (1980) advised that model building should incorporate both William of Ockam's 'Principle of Parsimony', which stated that 'one should not multiply causes without reason', and Bayesian philosophy. The use of Bayes' theorem allows for personal judgements about the relative merits of various alternative mechanisms in the model building process. Thus, whatever prior information is available should be used (Reilly, 1970).

This approach contrasts strongly with the school of thought that 'lets the data speak for themselves', as was advocated by Young (1983) in a discussion of time series and systems analyses of methods for evaluating dynamic, poorly defined systems. In accord with Popperian concepts of simplicity (Popper, 1959) the objective of the analysis was to find the simplest description of the data. In questioning the wisdom and scientific validity of attempts at complex computer-based model building, Young (1983) argued that the modeller should attempt, wherever possible, to construct a mathematical model whose size and complexity reflected not so much current perceptions of the system, but rather the information contained in the experimental data available from the system. It is clear that there is merit in both approaches, and, as is often the case with such polarized stances, it is likely that a satisfactory solution lies somewhere in the middle ground.

4.2 MODEL CATEGORIES

Physical, conceptual, mathematical and computer models of large and complex dynamic environmental systems are routinely constructed, generally falling into one of two categories. *Physical* models consisting of microcosms or model ecosystems attempt to isolate a representative segment of the environment within which the behaviour of a chemical is observed in order to predict environmental fate. *Mathematical* models attempt to define all the important processes acting on a chemical in the environment and incorporate them into a description of chemical behaviour as a function of the variables acting on the system.

The type of modelling approach adopted depends on the degree of accuracy and precision expected and the questions asked of the model. None the less, it is clear that if the environmental fate of many thousands of chemicals is to be determined, a mathematical rather than a physical model is the only practicable first step.

Mathematical models themselves may be categorized as *specific* or *evaluative*. Specific models describe the transport and degradation of a chemical in a particular situation, whereas evaluative models attempt to classify broadly the behaviour of chemicals in a hypothetical environment such as the 'Unit World'. Equilibrium partitioning models are evaluative and, using a minimum of information, seek to predict environmental distribution patterns for a compound and thereby identify environmental compartments of primary concern. When additional information is required in order to define the system more precisely, some of the usefulness and general applicability of evaluative models is lost.

The behaviour of a pollutant in the environment and its exposure evaluation may be described by a time-dependent method, which allows for spatially and temporally specified pollutant concentrations. In such an approach, provision must be made either for a comprehensive set of continuous environmental data measurements or, if a mathematical model is employed, an accurate compartmental configuration with particular transfer pathways. Alternatively, a time-independent method can be adopted, with less rigorous data requirements. In such as the exposure commitment method (Bennett, 1982), the objective is to determine the partitioning of pollutant amounts in transfer pathways and to estimate quantities reaching a designated receptor. The actual rates of transfer are comparatively unimportant.

The use of compartment mathematical models to describe the distribution and cycling of pollutants in ecosystems has been reviewed by Gydesen (1984), and was considered by the participants of a symposium on 'Modelling the Fate and Effect of Toxic Substances in the Environment' edited by Jörgensen (1984). The transport, distribution and toxic effects of polychlorinated biphenyls in ecosystems has been reviewed by Kalmaz and Kalmaz (1979).

4.3 MATERIAL BALANCE AND DIMENSIONAL ANALYSIS

When applying equations derived from compartmental analysis, the material balance of the system should be fulfilled as follows (Neely, 1980):

$$\begin{bmatrix}\text{Total material}\\ \text{in compartment}\end{bmatrix} = \begin{bmatrix}\text{Material added}\\ \text{(source)}\end{bmatrix} - \begin{bmatrix}\text{Material dissipated}\\ \text{(sink)}\end{bmatrix}$$

Thus, for a closed system the total mass is constant and at equilibrium the rate of change of the mass in the various compartments is zero.

To be dimensionally correct, an equation must obey the following rules (Riggs, 1963):

1. Quantities added or subtracted from one another must all have the same dimensions.
2. Quantities equal to each other must all have the same dimensions.
3. Any quantity may be multiplied by or divided by any other quantity without regard to dimensions. However, the resulting product or quotient must have appropriate dimensions so that rule 2 is not violated.
4. The dimensions of a physical entity are independent of its magnitude. Hence Δx and dx must have the same dimensions as x even though the differential dx is 'infinitesimally' small.
5. Pure numbers such as 'e' (the base of natural logarithms), etc. have no dimensions. Two important general classes of pure numbers of dimensionless quantities are the following:
   (a) exponents, including all logarithms;
   (b) ratios of two quantities with the same dimensions, e.g. trigonometric functions.
6. The dimensions of a quantity are not affected when it is multiplied or divided by a nondimensional quantity.

A chemical element mass balance (CEB) receptor model for Chicago has been used to determine the contributions from seven sources to the ambient total suspended (TSP) and respirable particulate (RP) matter (Scheff *et al.*, 1984). The CEB was applied as an independent analysis for 33 days of air measurements between July 1981 and February 1982. During this period, on average, soil-derived aerosol was found to make up the largest fraction of the TSP and RP at 33·5 and 7·3 $\mu g\, m^{-3}$, respectively. Other components were from limestone and cement sources, mobile sources, coal combustion, refuse incineration and steel industries.

The role of personal computers in environmental engineering has been highlighted with respect to a mass balance model of water quality conditions in Green Bay, Lake Michigan (Canale & Auer, 1987). Measuring 160 km by 22 km and with a mean depth of 15·8 m, the bay was divided into 19 control volumes or model cells. Twelve model cells were apportioned to surface cells exhibiting thermal stratification and seven model cells were allocated to the bottom. A mass balance computation was performed on each model cell for all variables of interest. These included chloride, total phosphorus, total organic carbon and dissolved oxygen. The mass balance included exchange between adjacent model cells, in terms of horizontal and vertical mass transport.

4.4 PARTITIONING MODELS

In its simplest form a partitioning model evaluates the distribution of a chemical between environmental compartments, based on the thermodynamics of the system. The chemical will interact with its environment and tend to reach an equilibrium state between compartments (McCall *et al.*, 1983*b*).

Rather than develop different models for the various phenomena occurring in the environment, partition models adopt a more general approach to the kinetics of transport, without reference to particular mechanisms such as diffusion, fluid flow and evaporation. Models consisting of compartmental systems best satisfy these requirements. Such compartments assume that various regions of the ecosystem can be represented by a series of 'ideal' volumes in which chemicals move from one volume to the next according to the laws of kinetics. These ideal volumes imply that all property variations are ignored and perfect mixing is assumed so that the outflow has the same properties as the contents of the compartment. Such basic assumptions restrict the applicability of compartmental models when considering ecological systems (Neely, 1980). Validation of a kinetic distribution model for chemicals in a standardized microcosm has been attempted by Figge *et al.* (1986). It was noticeable how the initial distribution of a chemical had a strong influence on the further development of its distribution pattern.

Unit world

Neely (1982) used a model 'Unit World' (Fig. 4.1) to represent the major environmental compartments. The volumes and physical properties of these compartments were selected to mimic a real environment as closely as possible. Transport processes continuously removing chemicals from the Unit World were presumed to be exit to the stratosphere and burial in the bottom sediments. The major transformation processes considered were hydrolysis, atmospheric oxidation, microbial degradation and aqueous photolysis. The model was operated by estimating the partition coefficients of the chemical between the various media:

$$C_a/C_w = H \text{ (Henry's constant)}$$

$$C_s/C_w = K_p \text{ (soil sorption coefficient)}$$

$$C_f/C_w = BF \text{ (bioconcentration factor)}$$

where C = concentration in water (w), soil (s), fish (f) and air (a).

These coefficients were estimated using the procedure summarized

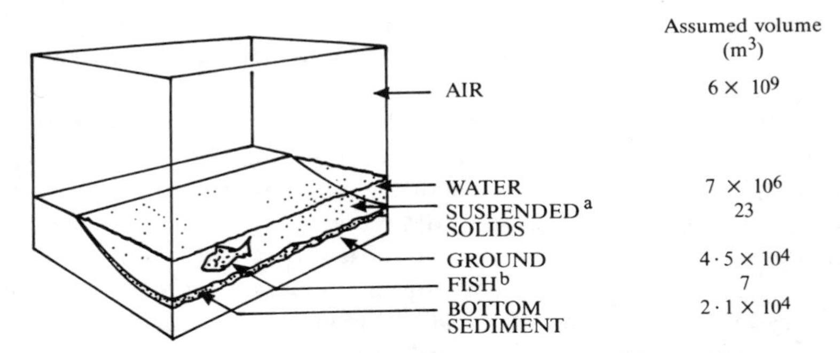

Fig. 4.1. A schematic representation of the 'Unit World' used to model the major environmental compartments. Reproduced from Neely (1982). [a] Suspended sediment is present at a concentration of 5 g m^{-3}; [b] Fish are present at a concentration of 1×10^{-6} g m^{-3}.

Table 4.1
Equations for estimating partition coefficients

Equation	*Reference*
$H = PM\ 16{\cdot}04/TS$	Dilling (1977)
$\log BF = 0{\cdot}85 \log K_{ow} - 0{\cdot}70$	Veith *et al.* (1979)
$K_P = \%$ organic[a] $(0{\cdot}6K_{ow})$	Karickhoff *et al.* (1979)
$\log K_{ow} = 6{\cdot}5 - 0{\cdot}89\ (\log S/M) - 0{\cdot}015$ (m.p.)	Banerjee *et al.* (1980)

where: P = vapour pressure (mm Hg)
M = molecular weight
T = absolute temperature
S = water solubility in g/m^3
K_{ow} = octanol/water partition coefficient
m.p. = melting point in °C
K_P = soil/water partition coefficient

[a] This is the amount of organic carbon in the soil, assumed to be 2% for ground and 40% for bottom sediments.
References selected by Neely (1982).

in Table 4.1. Given data for molecular weight, vapour pressure and water solubility, the environmental distribution of a chemical could be determined. The resulting pattern identified the compartment in which most degradation occurred. Therefore, for example, degradative reactions in the aquatic compartment could be ignored with a chemical deemed volatile.

Neely (1982) equated the reciprocal of the model residence time (a rate constant) to the assimilative capacity (AC) of the model world. Defining AC as the ability of an aquatic ecosystem to assimilate a substance without detriment to the ecosystem, Neely (1982) illustrated how AC can be used to calculate the maximum inputs to a system compatible with existing pollution control regulations.

The Unit World model was also used to investigate the water-contaminating potential of the 65 classes and 129 specific chemicals listed under Section 307(a) of the 1977 amendments to the Clean Water Law (Blair, 1981). Challenging the scientific validity of, for instance, listing chloroform and carbon tetrachloride separately from their class of halomethanes (Table 4.2), the Section 307 toxic pollutants were recategorized by chemical type as given in Table 4.3 (Callahan & Slimak, 1979). The model was presented with the properties of these chemicals necessary to generate a partitioning

Table 4.2
The 65 classes of toxic pollutants (in alphabetical order)

No.	*Chemical*	*No.*	*Chemical*
1	Acenaphthene (1)	34	Endrin (2)
2	Acrolein (1)	35	Ethylbenzene (1)
3	Acrylonitrile (1)	36	Fluoranthene (1)
4	Aldrin–dieldrin (2)	37	Haloethers (4)
5	Antimony (1)	38	Halomethane (8)
6	Arsenic (1)	39	Heptachlor (2)
7	Asbestos (1)	40	Hexachlorobutadiene (1)
8	Benzene (1)	41	Hexachlorocyclohexane (4)
9	Benzidine (1)	42	Hexachlorocyclopentadiene (1)
10	Beryllium (1)	43	Isophorone (1)
11	Cadmium (1)	44	Lead (1)
12	Carbon tetrachloride (1)	45	Mercury (1)
13	Chlordane (1)	46	Naphthalene (1)
14	Chlorinated benzenes (3)	47	Nickel (1)
15	Chlorinated ethanes (7)	48	Nitrobenzene (1)
16	Chlorinated naphthalene (1)	49	Nitrophenols (4)
17	Chlorinated phenol (2)	50	Nitrosamines (3)
18	Chloralkyl ethers (3)	51	PAH (13)
19	Chloroform (1)	52	PCB (7)
20	2-Chlorophenol (1)	53	Pentachlorophenol (1)
21	Chromium (1)	54	Phenol (1)
22	Copper (1)	55	Phthalate esters (6)
23	Cyanide (1)	56	Selenium (1)
24	DDT (3)	57	Silver (1)
25	Dichlorobenzene (3)	58	TCDD (1)
26	Dichlorobenzidine (1)	59	Tetrachloroethylene (1)
27	Dichloroethylene (2)	60	Thallium (1)
28	2,4-Dichlorophenol (1)	61	Toluene (1)
29	Dichloropropane (2)	62	Toxaphene (1)
30	2,4-Dimethylphenol (1)	63	Trichloroethylene (1)
31	Dinitrotoluene (2)	64	Vinyl chloride (1)
32	1,2-Diphenylhydrazine (1)	65	Zinc (1)
33	Endosulfan (3)		

Number in parentheses indicates number of chemicals in class. Total number of chemicals = 129.
Data from Neely (1982) (after Blair (1981)).

Table 4.3
Categorization of toxic pollutants by chemical type

Group	*Type*	
I	Pesticides	(9·19)[a]
II	Halogenated aliphatic hydrocarbons	(11·26)
III	Halogenated ethers	(2·7)
IV	Polychlorinated biphenyls	(1·7)
V	Monocyclic aromatic hydrocarbons	(14·23)
VI	Phthalate esters	(1·6)
VII	Polycyclic aromatic hydrocarbons	(5·17)
VIII	Nitrosamines	(1·3)
IX	Miscellaneous	(6·6)
X	Metals and inorganics	(15·15)
	Total	(65·129)

[a] Numbers in parentheses represent the EPA designation of number of classes and number of chemicals, i.e. there are nine classes of pesticides representing 19 different chemical species.
Data from Neely, 1982 (after Callahan & Slimak (1979)).

pattern showing the percentage of chemicals in air, water, bottom sediments and ground. The 15 metals and inorganics of Group X of Table 4.3 and 35 of the remaining 119 chemicals for which adequate data were unavailable were excluded from the preliminary profile. Table 4.4, from Neely (1982), presents the results of the analysis for those chemicals present in the water and sediment at a level of 10% (w/v), and shows that the Unit World model can be used to rank chemicals by their potential to contaminate water. Neely (1982) emphasized that the technique is only applicable to a set of well-defined organic structures. Analyses of inorganics, polymers or formulations were not believed possible.

OECD generic environment

Yoshida *et al.* (1983) proposed a simple equilibrium model for the prediction of the mass and concentration distribution fraction, mean residence time and concentration–time profile of a chemical released into the environment. The model environment chosen (Fig. 4.2) was the OECD generic environment, which is a hypothetical closed system. Hence the advective flow of air and water were not considered and characteristics such as volume, pH and temperature were fixed. The partition equilibria between environmental compartments in the

Table 4.4
Ranking of toxic pollutants by their potential to contaminate water

Chemical	*Category*	*Percentage in water and bottom sediments*
Bis (2-Cl-ethoxyl)methyl ether	III	98
Phenol	V	97·2
Dimethylphthalate	VI	97
2,4-Dinitrotoluene	V	95
2,4-Dimethylphenol	V	94
Isophorone	I	83
2,4-Dichlorophenol	V	80
Lindane	I	73
2-Cl-phenol	V	68
Nitrobenzene	V	68
Bis (2-Cl-ethyl) ether	III	69
Diethylphthalate	VI	63
Pentachlorophenol	V	53
4-Nitrophenol	V	52
Bis (2-ethylhexyl) phthalate	VI	52
Aldrin	I	52
DDT	I	52
Endrin	I	52
Heptachlor	I	52
Fluoranthene	VII	52
1,2,5,6-Dibenzanthracene	VII	51
1,12-Benzoperylene	VII	51
Chrysene	VII	51
3,4-Benzopyrene	VII	51
1,2-Benzanthracene	VII	51
Di-*n*-octyl phthalate	VI	51
Hexachlorobenzene	V	50·3
Chlordane	I	47
Acenaphthylene	VII	44
Fluorene	VII	39
Phenanthrene	VII	33
Acrolein	IX	26
2-Nitrophenol	V	24
Acrylonitrile	IX	23
4-Cl-phenyl phenyl ether	III	20
Bis (2-Cl)isopropyl ether	III	20
PCB	IV	18
Di-*n*-butyl phthalate	VI	14
Anthracene	VII	13
Naphthalene	VII	12·6
Bis(chloromethyl)ether	III	12
2-Cl-Naphthalene	VII	11
Endosulfan	I	10
2-Cl-ethyl vinyl ether	III	10

Data from Neely (1982).

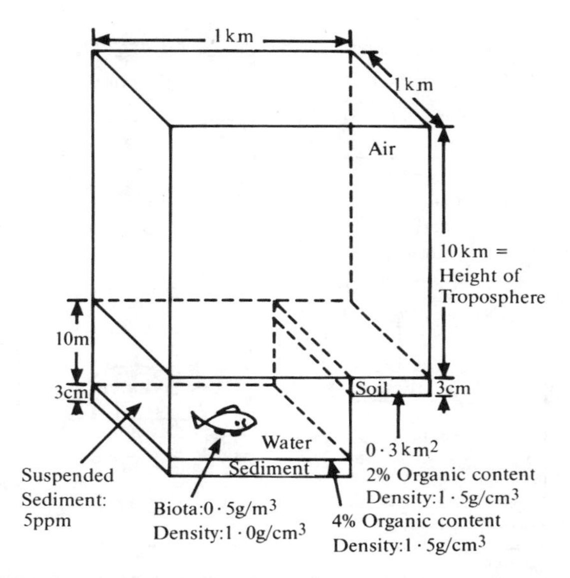

Fig. 4.2. OECD Generic Environment used in the equilibrium model (Working Party on Exposure Analysis, 1980). Reproduced from Yoshida *et al.* (1983).

model are shown schematically in Fig. 4.3. The equilibrium model assumed that transfer processes for chemicals between environmental compartments were rapid compared with their transformation processes, thus ensuring that chemicals achieved partition equilibrium. Clearly, such models cannot be applied to rapidly degradable chemicals whose transformation processes are relatively fast. Results from this model (Yoshida *et al.*, 1983) were consistent with data from the experimental addition of chlorpyrifos to a pond (Neely, 1979, 1980) and field data monitored by the Japanese Environment Agency (OHS, 1976–1980).

OECD environmental exposure analysis

The use of OECD premarket data in environmental exposure analysis for new chemicals has been reviewed by Hushon *et al.* (1983). The OECD Exposure Analysis Working Party (Umweltbundesamt, Berlin, December 1981) concluded that two expressions for environmental exposure should be considered when performing a first assessment of the potential hazard of a chemical, based on a minimum set of

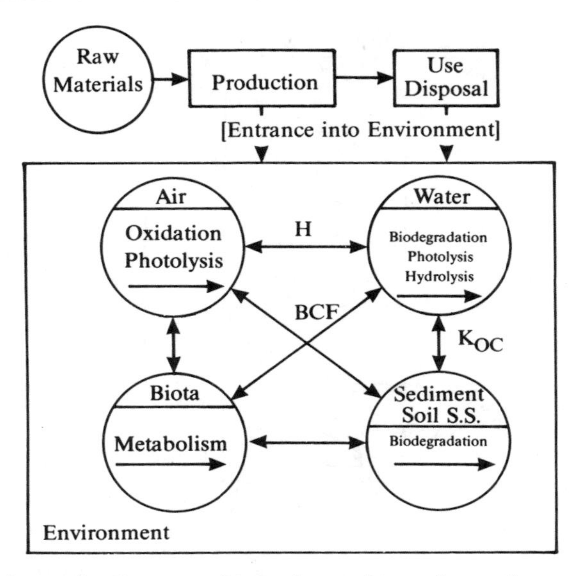

Fig. 4.3. Schematic diagram of transfer and transformation processes for chemicals in the environment. *H,* Henry's constant; *BCF,* biocentration factor in fish; K_{OC}, soil sorption coefficient. Reproduced from Yoshida *et al.* (1983).

premarketing data, namely:

Potential environmental distribution (PED)
Potential environmental concentration (PEC)

Both expressions allowed the estimation of exposure potentials of chemicals within environmental compartments of major concern. However, while PEC gives an estimation of the nonequilibrium (point source) situation for a discharged chemical within distinct ecosystems, the PED procedure estimates the widespread, equilibrium (steady-state) general environmental exposure (GEE) situation. Factors governing the latter category of exposure were believed to pertain to properties of the substance in the media normally obtainable from premarketing data elements (Hushon *et al.*, 1983). The resulting potential distribution was used along with available effects data to complete an initial hazard assessment (hazard ranking). Although a higher degree of reliability and accuracy for hazard assessment was possible from PEC estimations, which allowed for direct comparison with effects information, the extra specific data required often meant that PEC estimation was not feasible. Hushon *et al.* (1983) also observed that GEE is theoretically calculable by averaging an infinite number of geographically specific receptor sites.

For the estimation of environmental exposure, the OECD Exposure Analysis Group initially considered four models:

1. The *Klopffer Model* by Battelle Frankfurt for the Umweltbundesamt of the FRG, which presumed equilibria between media (Frische *et al.*, 1979*a,b*).
2. The *Neely Model*—based on laboratory and monitoring data for a pesticide in a pond environment (Neely, 1980).
3. The *Mackay Fugacity Model* in which fugacity capacities were used to explain various compartmental relationships, operating at several levels of sophistication depending on the data available (Mackay, 1979; Mackay & Paterson, 1981).
4. The *Wood Model,* developed for the US EPA, which was an adaptation of Mackay's Fugacity Model and also had several levels of complexity (Wood, 1981).

The Klopffer, Mackay and Wood models made slightly different assumptions of compartmental volume, but if degradation processes are ignored and chemical distribution is assumed to be at equilibrium, all reduce to essentially the same set of equations (Hushon *et al.*, 1983). 'Standard' compartmental volumes considered reasonable for evaluation purposes and considered suitable for default values were given previously in Fig. 4.2.

In the end, two models were excluded from the context of OECD hazard assessment: the Klopffer model because it made no provision for degradation or nonequilibrium processes, and the Neely model because, being designed for a specific experimental environment, it was considered inappropriate for use as a general screening tool for a broad range of chemicals. The two remaining, the Wood and Mackay models, are both based on the concept of fugacity and assume equilibrium or steady-state conditions between a set of environmental compartments. Essential physical and chemical data required for these two models which are, in their simpler versions, identical in predicting environmental compartmentalization, include the following:

1. Molecular weight.
2. Water solution.
3. Vapour pressure.
4. Soil sorption constant.
5. *n*-Octanol/water partition coefficient.

4.5 FUGACITY

A substance such as benzene, when at equilibrium, exists at different concentrations in air, water, sediment and other environmental phases or media. These equilibrium partitioning situations may be explained in fundamental thermodynamic terms. Gibbs showed that partitioning, or the maximization of the entropy of the system, can be expressed by equating the chemical potential of the substance in each phase. Comprehensively discussed by Prausnitz (1973), chemical potential has units of energy per mole and is conceptually elusive. Thus, Lewis (1901*a,b*) introduced the concept of fugacity as a more convenient descriptor of phase equilibrium.

Fugacity has units of pressure (Pa) and may be regarded as the 'escaping tendency' that a substance exerts from any given phase. The physical significance of fugacity is easily understood, since one molecule exerts 1/10 the escaping tendency of 10 (Mackay, 1979). Equilibrium is achieved between two phases when the 'escaping tendency' from one exactly matches that from the other. If the five environmental compartments, atmosphere (A), soil (B), a lake (C), sediment (D) and aquatic biota (E), are considered as phases:

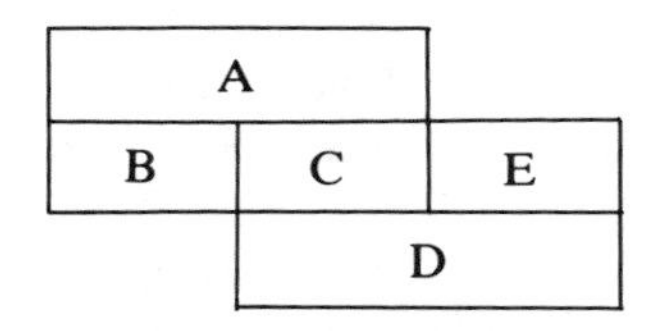

then at equilibrium, fugacity (f) may be mathematically expressed as

$$f_A = f_B = f_C = f_D = f_E$$

Fugacity may be considered as being to mass diffusion what temperature is to heat diffusion. Heat always diffuses from high to low temperature, and mass always diffuses from high to low fugacity. Diffusion directions are more clear when expressed in atmospheres (or °C) rather than concentrations such as μmol m^{-3} (or cal m^{-3}). Just as temperatures (°C) can be related to heat concentrations (cal m^{-3}), using a proportionality constant, giving heat capacity as cal/(m^3 °C), so fugacity can be related to concentration (C) using a fugacity capacity constant (Z) with units of mol m^{-3} atm, as

$$C = Zf$$

Thus, if

$$f_A = f_B$$

it follows that

$$C_A/Z_A = C_B/Z_B$$

or

$$C_A/C_B = Z_A/Z_B = K_{AB}$$

where K_{AB} is the dimensionless partition coefficient controlling the distribution of the substance between the two phases A and B and is, in effect, the ratio of the fugacity capacities. Mackay and Paterson (1981) applauded the elegance of this approach, which they illustrated by considering a 10-phase system in which there were potentially 90 definable partition coefficients that were constrained in value with respect to each other such that only nine could be independently defined. With only 10 fugacity capacities, the 90 partition coefficients were therefore merely all possible ratios of those 10 values.

The value of Z will depend on temperature, pressure, the nature of the substance and the medium in which it is present. Its concentration dependence is usually slight in very dilute situations. The physical significance of Z is that it quantifies the capacity of the phase for fugacity. At a given fugacity, if Z is low, C is low and only a small amount of substance is necessary to exert the escaping tendency. Toxic substances thus tend to accumulate in phases where Z is high, or high concentrations can be reached without creating high fugacities. Hydrophobic organics tend to partition into lipid phases, because that is where their Z value is at a maximum (Mackay & Paterson, 1981).

Environmental applications

Since location-specific environments are highly complex, the pragmatic approach is to use a general or 'evaluative' model of the hypothetical, but typical, environment such as was described by Baughman and Lassiter (1978) and developed by Smith *et al.* (1977/8). In the evaluative model, no attempt is made to simulate the real environment. Instead, an attempt is made to provide behavioural information characteristic of the substances' relative partitioning into each environmental phase, the dominant reactions, the principal intercompartmental transport processes and the overall persistence. It is rarely possible to calculate actual environmental concentrations since these vary both spatially and temporally. The model merely provides a structure within which the available data can be systematically processed by the use of equations that have inherent physical validity.

Mackay (1979) indicated two general areas in which fugacity–concentration calculations can contribute to a better understanding of the fate of toxic substances. In the first application, concentration data from an existing multiphasic pollutant monitoring programme can be converted to fugacities and the fugacity levels compared. The second application is concerned with the probable environmental fate of a compound that is being marketed for the first time.

4.5.1 Fugacity capacities (*Z*)

Vapour phase or the atmosphere

In the vapour phase, the fugacity is rigorously expressed by

$$f = y\phi P_{\mathrm{T}} \approx P$$

where y is the solute mole fraction (solute is truly gaseous), P_{T} is the total pressure (i.e. atmospheric pressure, in Pa), ϕ is the fugacity coefficient (dimensionless) and P is the partial pressure.

The fugacity coefficient is used to account for nonideal behaviour but, with the exception of solutes such as carboxylic acids that associate in the vapour phase, is close to unity at atmospheric pressure. Concentration (C) is related to partial pressure from the gas law, thus

$$C = n/v = P/RT = f/RT - fZ$$

where n represents moles and v volume.

Thus, Z_{a} for vapours is equivalent to $1/RT$ ($4{\cdot}04 \times 10^{-4}$ mol m^{-3} Pa at 25°C); Z_{a} is independent of the nature of the substance, but is dependent on temperature.

Liquid phase or water bodies

In aqueous solution, the fugacity is given by

$$f = x\gamma P^{\mathrm{s}}$$

where x is the mole fraction, P^{s} is the vapour pressure of solute at system temperature (pure liquid) and γ is the liquid-phase activity coefficient (Raoult's law).

Generally, for nonionizing substances, γ increases to an 'infinite dilution' as x tends to zero. Mackay and Paterson (1981) expressed this relationship as

$$\ln \gamma = K(1-x)^2$$

Since in most environmental situations x is negligible, $\ln \gamma$ is approximately equal to K, the partition coefficient.

For infinite dilutions, Z may be equated to f and C by

$$Z = C/f = C/P = 1/H = x/v_w f = 1/v_w \phi P^s$$

where v_w is the molar volume of the solution ($m^3\ mol^{-1}$).

Fugacity or partial pressure (P) is usually related to concentration by the Henry's law constant (H) as

$$P = HC$$

It follows that for water, Z_w is simply $1/H$; H is calculated as the ratio of pure substance saturated vapour pressure (P^s) to solubility (C^s). Bounds to the equations for solution fugacity are as follows:

(a) The solution must be in dissolved form at a concentration less than or equal to saturation, and not present environmentally in a sorbed form associated with suspended mineral or organic matter. Polychlorinated biphenyls (PCBs) or polynuclear aromatic hydrocarbons (PAHs) by virtue of their hydrophobicity may be present in colloidal form at concentrations in excess of their solubility. Such forms of solute must be excluded from calculations of solution fugacity.

(b) Pure solutes may undergo phase transitions (melting, boiling, crystallization) at environmental temperatures. Published data on say, PCB solubilities, usually refer to the solid pure isomers, whereas the vapour pressures may refer to supercooled liquid mixtures. For calculations of H, both P^s and C^s must refer to the same phase.

(c) Occasionally γ may be calculated from molecular structure, and providing that data are available for P^s, H can be deduced as $v\gamma P^s$. This calculation has been discussed further by Yalkowsky (1979), Yalkowsky and Valvani (1979), Valvani and Yalkowsky (1980) and Mackay *et al.* (1980).

Sediment or suspended sorbed phases

Sorption equilibria are usually expressed as equations or isotherms relating dissolved to sorbed concentrations. For most hydrophobic compounds in dilute concentrations, Karickhoff *et al.* (1979) suggested the linear equation:

$$X = K_P C$$

where X is the sorbed concentration (moles of solute per 10^6 g sorbent, wet or dry; alternatively $\mu g\, g^{-1}$), K_P is the sorption coefficient (m^3 water per 10^6 g sorbent) and C is the concentration ($g\, m^{-3}$ or mg $litre^{-1}$).

The sorbent concentration expressed as volume fraction is termed S. If density is ρ, then concentration is expressed as $S\rho$ (typically $10^{-5}\, g\, ml^{-1}$ or 10 mg $litre^{-1}$). Data are conveniently recorded as $g\, m^{-3}$ or $10^6 S\rho$. The concentration of sorbed material (C_s) ($mol\, m^{-3}$ sorbent) is thus $X\rho$ $mol\, m^{-3}$ (Mackay & Paterson, 1981).

At equilibrium, the fugacities of the sorbed and dissolved material are equal, therefore

$$f = HC = C_s/Z_s$$

where Z_s is the sorbed-phase fugacity capacity. It follows that

$$Z_s = C_s/HC = X\rho/H(X/K_P) = K_P/H$$

where $K_P\rho$ is the dimensionless partition coefficient (moles/unit volume ratio).

If 1 m^3 of solution contains a low concentration of sorbent of volume fraction S, then the concentration of dissolved material is $Z_w f$ ($= f/H$ $mol\, m^{-3}$). The sorbed concentration is fZ_s or C_s $mol\, m^{-3}$ of sorbent or a total of fZ_sS mol or $fK_P\rho S/H$. Therefore

$$\text{Total amount} = f/H + fK_P\rho S/H = f(1 + K_P\rho S)/H$$

$$\text{Fraction dissolved} = (f/H)/f(1 + K_P\rho S/H = 1/(1 + K_P\rho S)$$

$$\text{Fraction sorbed} = (fK_P\rho S/H)/f(1 + K_P\rho S)/H$$

$$= K_P\rho S/(1 + K_P\rho S)$$

Conventionally, the sorbent concentration is expressed as S' $g\, m^{-3}$ or mg $litre^{-1}$ ($= 10^6 S$). Thus, the group $K_P\rho S$ is replaced by the group $10^{-6} K_P S'$.

The analysis may be extended to sorption on atmospheric particles. However, the sorbed solute must not be physically trapped in or enveloped by the particle, as may occur with PAHs formed during combustion and associated with soot particles.

Octanol phase

Mackay and Paterson (1981) suggested that

$$Z_o = 1/v_o \gamma_o P^s$$

where Z_o represented Z for octanol, γ_o was the activity coefficient of the solute in octanol, P^s the vapour pressure of the liquid solute (at the system temperature) and v_o was the molar volume of octanol saturated with water.

When fugacities are equal, the octanol–water partition coefficient (K_{ow}) is expressed as a concentration ratio (mol m^{-3} or g m^{-3}), thus

$$K_{ow} = C_o/C_w = Z_o/Z_w = v_w \gamma_w / v_o \gamma_o$$

The vapour pressures cancel each other, v_w and v_o are constants, as is γ_o within the range 1–10. It is clear then, that the value of K_{ow} is determined mainly by γ_w. Since γ_w controls aqueous solubility, Mackay *et al.* (1980*a*) developed the following expressions relating solubility with K_{ow}:

For liquids: $\ln K_{ow} = 7{\cdot}494 - \ln C^s$

For solids: $\ln K_{ow} = 7{\cdot}494 - \ln C^s + 6{\cdot}79(1 - T_m/T)$

Biotic phases

If the biota are regarded as part octanol (fraction y) and the volume fraction of biota is B, then

$$Z = ByK_{ow}/H$$

where K_{ow} is the octanol–water partition coefficient.

When the wet weight bioconcentration factor K_B is used instead of the partition coefficient it is considered to be identical to K_P. Expressed on a wet volume basis, it is equivalent to the group $K_P\rho$, where ρ is the biota density. As ρ is approximately unity, differences between the K_P and $K_P\rho$ are dimensional only. A modified bioconcentration factor is employed in calculations based on dry weight or lipid weight. Such bioconcentration factors have been discussed further by Neely *et al.* (1974) and Veith *et al.* (1979*b*).

Pure solid and liquid phases

For pure solids or liquids, fugacity is equivalent to the vapour pressure (P^s). As a substance's concentration (C) is the inverse of the molar volume v_s ($m^3\,mol^{-1}$), Z_P is given by

$$Z_P = C/f = 1/P^s v_s$$

Pure phases occur environmentally only when the solute solubility in a given phase is exceeded and the solute precipitates out.

Since

$$K_{aP} = Z_a/Z_P = P^s v_s/RT$$

It can be seen that vapour pressure is described by the air-pure solute partition coefficient and water solubility C^s is essentially the water-pure solute partition coefficient:

$$K_{wP} = Z_w/Z_P = P^s v_s/H = C^s v_s$$

The solubility (C^s) is defined as follows:

For liquid solutes: $C^s = 1/v_w \gamma_w$

For solid solutes: $C^s = P^s_s/P^s_L V_w \gamma_w$

where P^s_s is the solid vapour pressure and P^s_L is the liquid vapour pressure.

Inert solutes have near-infinite Z values in their pure solute form because of very low values of P^s. Examples of these sparingly soluble substances are ceramics, polymers and some metal salts.

Relationships between fugacity capacities and partition coefficients are illustrated in Fig. 4.4 and summarized fugacity capacity definitions are presented in Table 4.5. The acquisition of such data as solubility and vapour pressure is a prerequisite to the determination of Z values for a chemical in stated environmental phases.

4.5.2 Fugacity models

Calculations of environmental partitioning depend upon assumptions of media volume and solute quantities. As any concentration can be obtained by judicious juggling of media volumes and solute amount, only the relative concentrations of a substance between phases should have significance. The selection of suitable volumes for each environmental compartment is critical, since the mass distribution of the substance depends on the assumed phase volumes. In general, fugacity models seeking to predict the environmental distribution of a substance assume a model 'Unit World' (Mackay & Paterson, 1981) consisting of 1 km^2 of land area with soil, some water, sediment and biota, and with atmosphere above. Deep soils or deep oceanic waters not accessible over a period of say, one year, are not considered.

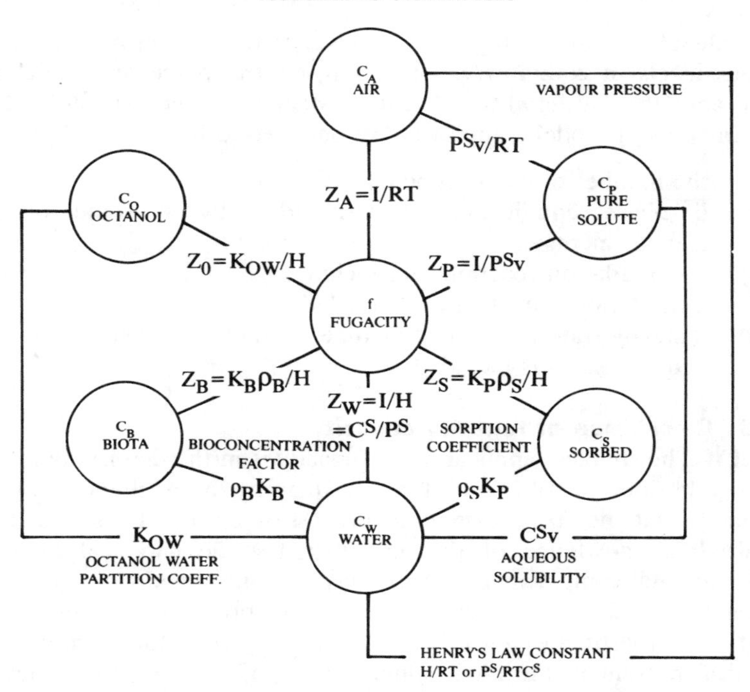

Fig. 4.4. Relationships between fugacity capacities and partition coefficients (symbols are defined in Table 4.5). Reproduced from Mackay *et al.* (1983*a*).

Table 4.5
Definition of fugacity capacities

Compartment	*Definition of Z* ($mol\,m^{-3}\,Pa$)
Air	$1/RT$ $R = 8{\cdot}314\ \mathrm{Pa\,m^3\,mol^{-1}\,K}$ T = Temp. (K)
Water	$1/H$ or C^s/P^sC^s = aqueous solubility ($\mathrm{mol\,m^{-3}}$) P^s = vapour pressure (Pa) H = Henry's law constant ($\mathrm{Pa\,m^3\,mol^{-1}}$)
Solid sorbent (e.g. soil sediment, particles)	$K_P\rho_s/H$ K_P = partition coeff. (litre $\mathrm{kg^{-1}}$) ρ_s = density (kg $\mathrm{litre^{-1}}$)
Biota	$K_B\rho_B/^H$ K_B = bioconcentration factor (litre $\mathrm{kg^{-1}}$) ρ_B = density (kg $\mathrm{litre^{-1}}$)
Pure solute	$1/Psv$ v = solute molar volume ($\mathrm{m^3\,mol^{-1}}$)

Data from Mackay *et al.* (1983*a*).

A valuable feature of fugacity models is that they can be applied at various levels of complexity, depending on the perceived modelling need and the availability of data (Mackay *et al.*, 1983*a*). The determinants of model complexity are believed to be

(a) the number of compartments considered;
(b) if phase equilibrium is assumed between some or all compartments;
(c) if degradation reactions are included;
(d) if advection processes are included;
(e) if steady state is assumed or time dependence of concentration and emissions is included.

4.5.3 Calculation of fugacity capacity

Level I: The simplest application of fugacity is in the determination of the equilibrium distribution of a set quantity of a chemical. It is assumed that no transformation processes occur. Thus, what is required is knowledge of the Z values for the chemical in each environmental compartment. These may be determined from chemical parameters and sorption characteristics as previously outlined. In addition, an estimate must be made of the sum total amount of chemical present in the environment at any given time (M moles of solute). Assigning each compartment a subscript i, at equilibrium (Mackay *et al.*, 1983*a*):

$$f_1 = f_2 = f_3 = \cdots f_i$$

and

$$M = \sum C_i V_i = \sum (f_i V_i Z_i) = f_i \sum (V_i Z_i)$$

therefore

$$f_i = M \Big/ \sum (V_i Z_i)$$

and

$$M_i = f_i V_i Z_i$$

where M_i is the number of moles in each compartment ($\sum M_i = M$), V_i is the accessible compartment volume (m^3) and C_i is the concentration in each phase.

Using the data in Table 4.6 for a trichlorobiphenyl, Fig. 4.5 illustrates a simple Level I fugacity calculation for a five-compartment

Table 4.6
Physicochemical properties of a trichlorobiphenyl (TCB)

Properties	*Air*	*Water*	*Sediment*	*Suspended sediment*	*Biota*
Partition coefficients ($1\,kg^{-1}$)			Kp_s 9 530	Kp_P 95 300	K_B 28 800
Fugacity capacities ($mol\,m^{-3}\,Pa$)	Z_A 0·0004	Z_w 0·013	Z_s 186	Z_P 1 860	Z_B 375
Transport parameters ($mol\,Pa^{-1}\,h^{-1}$)		D_{Aw} 64·1	D_{ws} 13·0	D_{wP} 313	D_{wB} 6·24
Reaction rate constants (h^{-1})		k_w $1{\cdot}0 \times 10^{-5}$	k_s $3{\cdot}5 \times 10^{-5}$		

Molecular weight = 257 ($g\,mol^{-1}$).
Henry's constant (H) = 77 ($Pa\,m^3\,mol^{-1}$).
$\log K_{ow} = 5{\cdot}76$.
Data from Mackay *et al.* (1983*a*).

system. The air–water transfer rate coefficient selected by Mackay *et al.* (1983*a*) is lower than that generally used, in order to reconcile observed air and water concentration and mass balances as discussed by Mackay *et al.* (1983*b*).

This calculation gives rise to two different classes of information. It identifies where most of the solute partitions, in this case sediment (56·9%), followed by air (24·6%), water (10·6%) and suspended sediment (7·58%), are to be found. In addition, the calculation shows where the highest concentrations occur; here they are in fish, where concentrations are some 30 000 times greater than in water. The absolute concentrations are not significant since they depend on the assumed amount of chemical and compartmental volume. However, care should be taken to avoid concluding that because significant PCB bioconcentration occurs in fish, that is where most PCBs partition.

Level II: In this calculation, advection and reactions such as photolysis, hydrolysis, oxidation and biodegradation are added to the Level I equilibrium and steady-state system. A Level II calculation is useful from a regulatory point of view, since by knowing the proportions of a substance advected into the system and the proportion derived from local sources, it is possible to calculate the relative

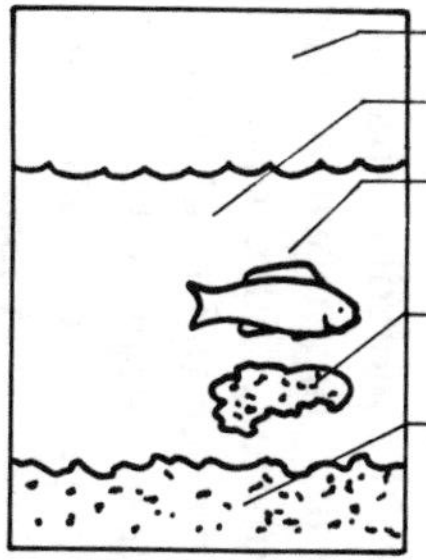

	VOLUME	Z	VZ	C=fZ	m=CV	
	m^3	mol m^{-3}Pa		mol m^{-3}	mol	%
AIR	$6\cdot0\times10^9$	$4\cdot0\times10^{-4}$	$2\cdot4\times10^6$	$4\cdot1\times10^{-11}$($10\cdot5$ng m^{-3})	0·246	24·6
WATER	80×10^6	0·013	$1\cdot04\times10^6$	$1\cdot3\times10^{-9}$($0\cdot34$ng l^{-1})	0·106	10·6
FISH	80	375	$0\cdot03\times10^6$	$3\cdot9\times10^{-5}$($0\cdot01$μg g^{-1})	0·0031	0·31
SUSPENDED SEDIMENT	400	1860	$0\cdot74\times10^6$	$1\cdot9\times10^{-4}$($0\cdot033$μg g^{-1})	0·076	7·58
BOTTOM SEDIMENT	3000	186	$5\cdot58\times10^6$	$1\cdot9\times10^{-5}$($0\cdot003$μg g^{-1})	0·569	56·9
TOTAL			$9\cdot79\times10^6$		1·000	100

$$f = M/\Sigma VZ = 1\cdot0/0/9\cdot79\times10^6 = 1\cdot02\times10^{-7}\ Pa$$

Fig. 4.5. Level I calculation for a trichlorobiphenyl, illustrating equilibrium distribution with no reaction of one mole of chemical. Reproduced from Mackay *et al.* (1983*a*).

effects of reducing local or external emissions (Mackay & Paterson, 1981).

Level III: This calculates the nonequilibrium distribution of a substance. Incorporated into the calculation are the steady-state input, transformations and intercompartmental transfers. The transfer rate between compartments is described by an exchange rate constant that is determined by the difference in fugacity. The steady-state transfer rate (N) with units mol h^{-1} is defined by the equation:

$$N = D_{12}(f_1 - f_2)$$

where 1 and 2 are two compartments; D has units of mol h^{-1} atm^{-1}.

The solute may be introduced into one or more compartments at a rate of I_i mol h^{-1}, therefore at steady-state for compartment i:

$$I_i = V_i C_i K_i + \sum_j D_{ij}(f_i - f_j)$$

The summation is over all compartments except i. Rearranging, one obtains

$$I_i = f_i\left(V_i K_i Z_i + \sum_j D_{ij}\right) - \sum_i (D_{ij} f_j)$$

Therefore, with n compartments this reduces to n simultaneous linear equations with (n) unknowns (f_i). Transformation, transfer rates and C_i can then be calculated by solving for f_i.

The Level III calculation illustrated in Fig. 4.6 shows that constraint from the air–water volatilization rate constant reduces air advective loss to 1·03 m mol h^{-1}, thereby increasing the relative importance of other reaction processes. Mackay *et al.* (1983*a*) suggested that a Level III calculation is most appropriate and particularly enlightening for regulatory purposes. This statement is supported by calculations similar to those in Fig. 4.6, but for mono-, di-, tri- and tetra-chlorobiphenyls. Congeners with few chlorines are fairly short-lived, have a reduced tendency to partition into sediments and biota, and most reactions occur in the water column, advection with air and burial in sediment being relatively insignificant. However, as chlorine number increases, amounts and persistence increase, as does the proportion of substance partitioned into sediments and biota. At the same time, sediment degradation and advection supersede and eventually dominate water column degradation as parameters of environmental significance. Thus, the more persistent higher chlorine congeners emerge as the most hazardous chemical group.

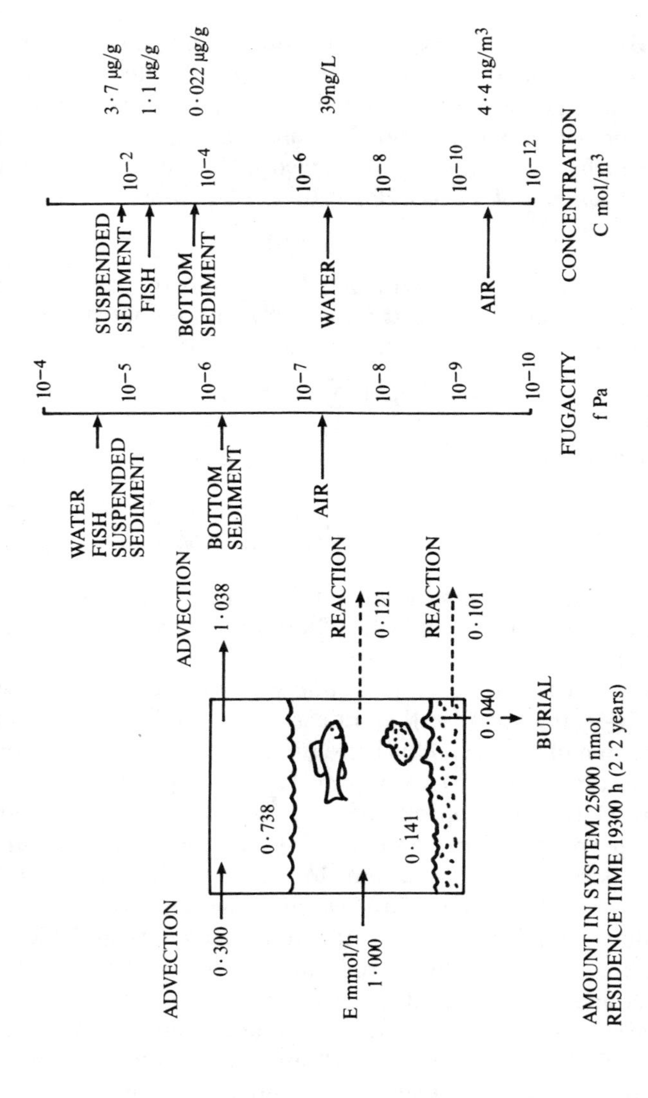

Fig. 4.6. Level III calculation for a trichlorobiphenyl. Reproduced from Mackay *et al.* (1983*a*).

Level IV: This level and the Exposure Analysis Modelling System (EXAMS) (Burns *et al.*, 1982) are nonequilibrium, with reaction and advection being in an unsteady state. Mathematically, Level IV can be expressed as a set of differential equations, such that

$$I_i(t) = f_i\left(V_i K_i Z_i + \sum_i D_{ij}\right) - \sum_i D_{ij} f_i + V_i Z_i (\mathrm{d}f_i/\mathrm{d}t)$$

where t is time.

Given the initial condition and appropriate input rates, change in concentration with time can be calculated for the various phases. This type of calculation can identify how long it will take for a substance to build up to hazardous levels and will also determine its persistence after emissions have ceased. Figure 4.7 shows the build up in concentrations and fugacity to the steady-state values given in the Level III calculation (Fig. 4.6), and then shows the subsequent decline after emissions are reduced. From Fig. 4.7 it can be seen that in this instance sediments are slowest to accumulate and slowest to lose a substance once incorporated. The figure shows that it would take 15 years to achieve a reduction of one order of magnitude in sediment concentration.

A Quantitative Water, Air, Sediment Interaction (QWASI) fugacity model describing the fate of chemicals in lakes (Mackay *et al.*, 1983*b*) has also been extended to treat rivers (Mackay *et al.*, 1983*c*). In the model, all rates were expressed as products of fugacity (f) and a transport or transformation parameter (D). The latter authors identified two major approaches in the quantification of river processes. These were the *sectioning* approach in which the river is treated as a series of connected lakes or volumes, each of which is described by a set of equations incorporating inflow and outflow, and the *analytical* approach. In the latter, integrated differential equations describe the changes in composition of the water and sediment continuously with flow-time or location. Within either the sectioning or the analytical approach, steady or unsteady conditions may be assumed.

Details of mass balance and other equations can be found in the worked example of a QWASI fugacity model for describing the fate of chemicals in rivers given by Mackay *et al.* (1983*c*). Phenol degradation has been simulated by the EXAMS model and field tested in the Monongahela river, Pennsylvania, forming part of the effluent from a steel plant (Pollard & Hern, 1985). A model describing rates of transfer processes of organic chemicals between atmosphere and water

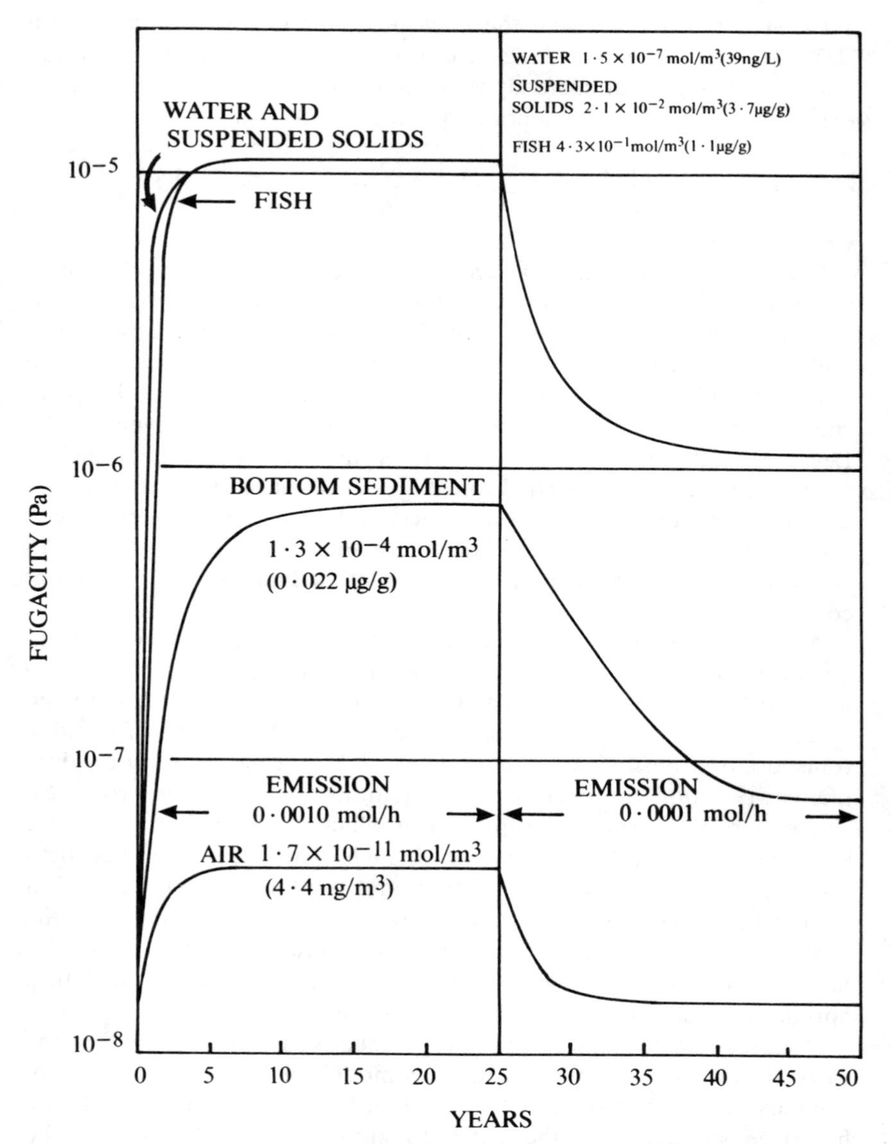

Fig. 4.7. Level IV calculation for a trichlorobiphenyl showing response to emission changes. Reproduced from Mackay *et al.* (1983*a*).

was reported by Mackay and Paterson (1984) and Mackay *et al.* (1986). The fugacity concept has also been used to describe the uptake of organic chemicals from water by fish (Mackay & Hughes, 1984).

Concentration variations within an environmental compartment have been discussed in general terms by Mackay and Paterson (1984) and reviewed for air by Georgopoulos and Seinfeld (1982) and for water by Dean (1981). Interest in the heterogeneous spatial distribution of a chemical, using a probability density function, arises from a concern for concentration 'peaks'. For instance, a mean concentration value is of use in some general modelling applications, but for toxicological purposes it is often necessary to know if some fraction of the environment experiences a hazardous concentration, even though overall mean values are considered 'safe'.

4.6 SYSTEM DESIGN

Blau and Neely (1983) identified four major components in every ecosystem undergoing perturbation by an outside influence. The first of these components is a source of perturbation, usually represented as the input dosage schedule. Second are the intercompartmental mass transport processes. Kinetic processes such as biodegradation represent the third major component. Then, completing the mass balance, are the outputs to suitable environmental sinks. Obeying their own adjuration to keep the model simple, adding component parts only as required, the latter authors outlined ecosystem components suitable for separate study. These are presented in Fig. 4.8. Blau and Neely (1983) also recommended that the questions needing to be answered be kept firmly in mind. Thus, as the questions change so too should the model.

The three categories of data flow (materials balance, environmental parameters and chemical properties) accepted by multimedia models proposed by Eschenroeder (1981) provide a decision procedure for an optimal assemblage of compartmental models and transfer processes. Decisions are based on preliminary dynamic partitioning analyses indicating gross trends. From these, a model can be constructed that addresses the main issues of the specific assessment being undertaken. The system design underlying multimedia modelling is outlined in Fig. 4.9 and includes an option to bypass the decision process and use all compartmental models and transfer processes at the outset.

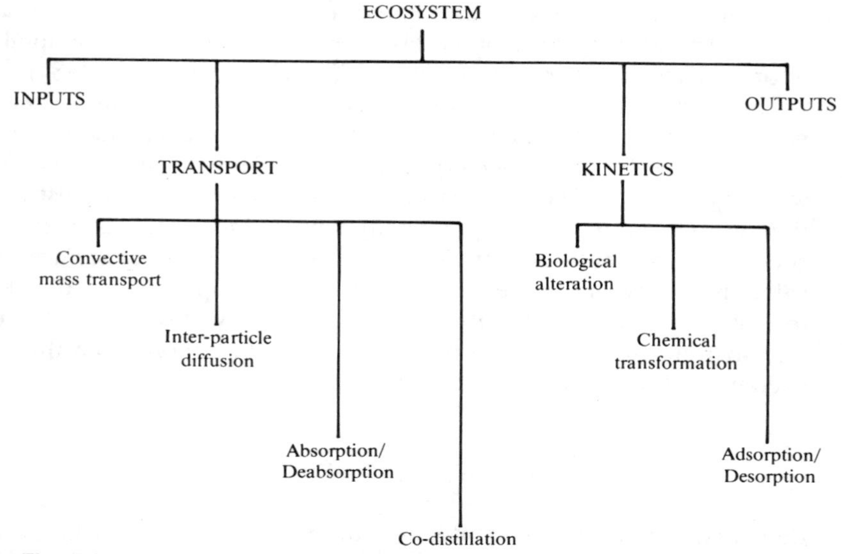

Fig. 4.8. Ecosystem components. Data from Blau and Neely (1983).

4.7 MODEL EVALUATION

Model evaluation is concerned with identification and quantification of the uncertainties associated with predictions from environmental assessment models. Such knowledge is vital if decisions as to a particular course of action are to be based on data from predictive models. Because of the diversity of existing environmental fate models, no one method of evaluation is adequate for the task of evaluation, but Shaeffer (1980) proposed a systematic methodology applicable to most assessment models. This strategy may be viewed as incorporating six tasks:

1. Model examination.
2. Algorithm examination.
3. Data evaluation.
4. Sensitivity analyses.
5. Validation.
6. Code comparisons.

Model examination determines whether a model adequately represents the phenomena, processes and actions of interest to the model

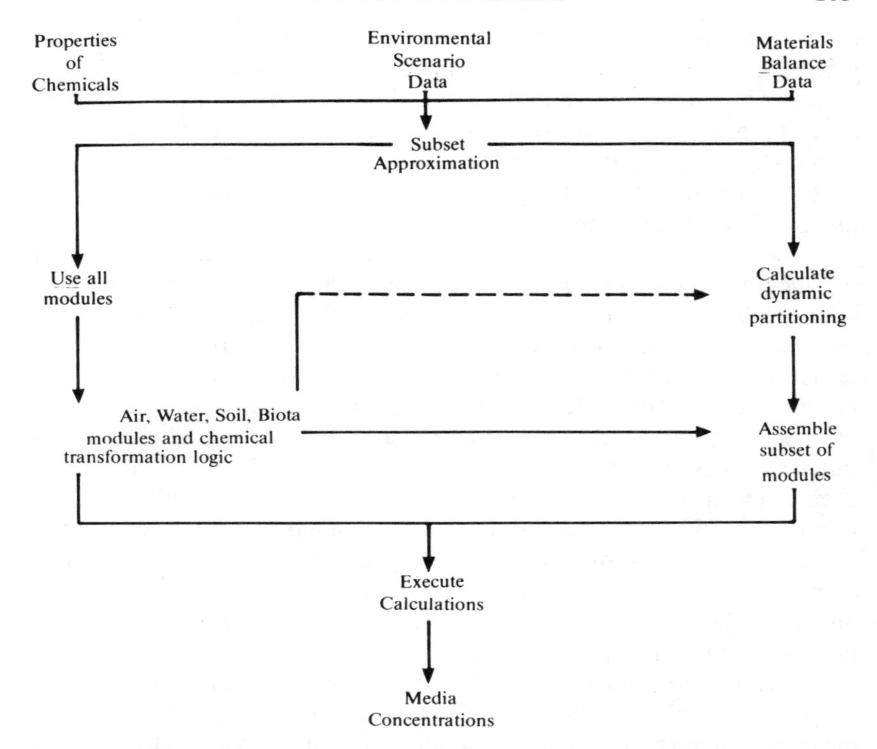

Fig. 4.9. System design underlying multimedia modelling. Reproduced from Eschenroeder (1981).

user. Model examination is primarily concerned with evaluation within a range of model applicability defined by the model documentation and its associated computer encodement. Thus, the behaviour of the model when subjected to abnormal input conditions is not considered.

Algorithm examination questions the numerical techniques selected for the model computer code and ensures that the numerical solutions obtained are unique.

Data evaluation determines both the quality of data input to the computer codes and comparison data with which model predictions are validated. Data quality is expressed in terms of accuracy (systematic error present), precision (random error present) and completeness. Where there are insufficient data to develop a probability density function (PDF) and insufficient information about the characteristics of an environmental parameter to permit the adoption of any particular PDF, then the method of Tiwari and Hobbie (1976) is useful. This procedure, based on the principle of maximum entropy, leads to a PDF that agrees with the available data, but which is maximally noncommital to missing information (Shaeffer, 1980).

Sensitivity analysis seeks to identify those key parameters for which the greatest accuracy and precision are needed to reduce uncertainty in model predictions. If the exact mathematical relationship between the model inputs and outputs are known, then an analytic approach such as Perturbation Theory (Tomovic, 1963) may be applied to sensitivity analyses. If an analytic approach is not possible, then a statistical or nonstatistical numerical approach can be used. Typical is the nonstatistical procedure of varying the value of an input parameter by a given percentage about a nominal value while maintaining all other parameters constant (Miller *et al.*, 1976). Values of a sensitivity coefficient may then be calculated, expressing the sensitivity of an output quantity to each input quantity. Spearman's nonparametric partial rank correlation coefficient (Siegel, 1956; Conover, 1971) may be used to remove the effects of all but one input parameter on a given output variable even though all input parameters are varied simultaneously from one run to another.

Validation procedures depend on the nature of output data. If the output is statistical, then the accuracy and precision of predicted and measured quantities should be similar (Shaeffer, 1980). However, if insufficient data are available to characterize the output as statistical, a test of derivative validity might be applied, where deviation of a predicted value from a measured value is evaluated quantitatively in terms of a deviation coefficient or quantitatively on a subjective validity scale. Alternatively, a variety of models might be compared with each other to establish 'comparative validity' (Goodall, 1972).

Such tests are applicable only to single output variables. When the model has several output parameters, in any given situation one might pass a test of validity and yet the others might fail. To test the 'global validity' of a model, each output quantity must be separately identified as being valid. It is also possible that a model may be proved valid for one set of input parameters and invalid for another, due to the inclusion of incorrect physical assumptions or refuted theories. Meticulous model examination would, however, help to reduce the likelihood of such occurrences.

Code comparison enables the user to select those most suitable for a

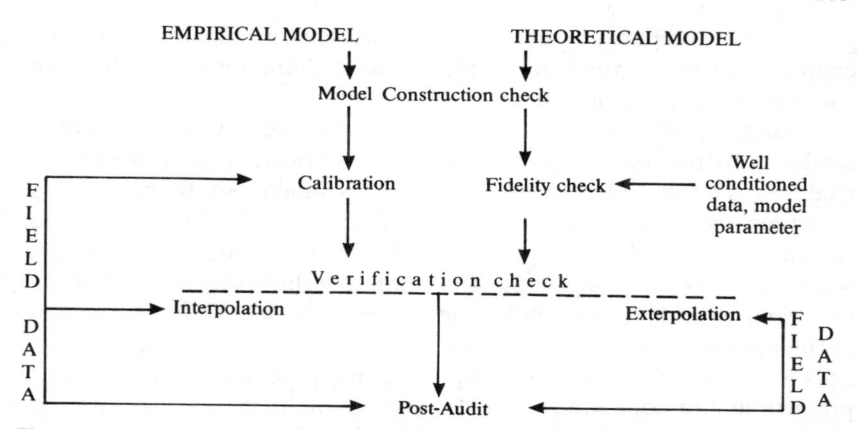

Fig. 4.10. The process of model evaluation. Reproduced from Dickson *et al.* (1982).

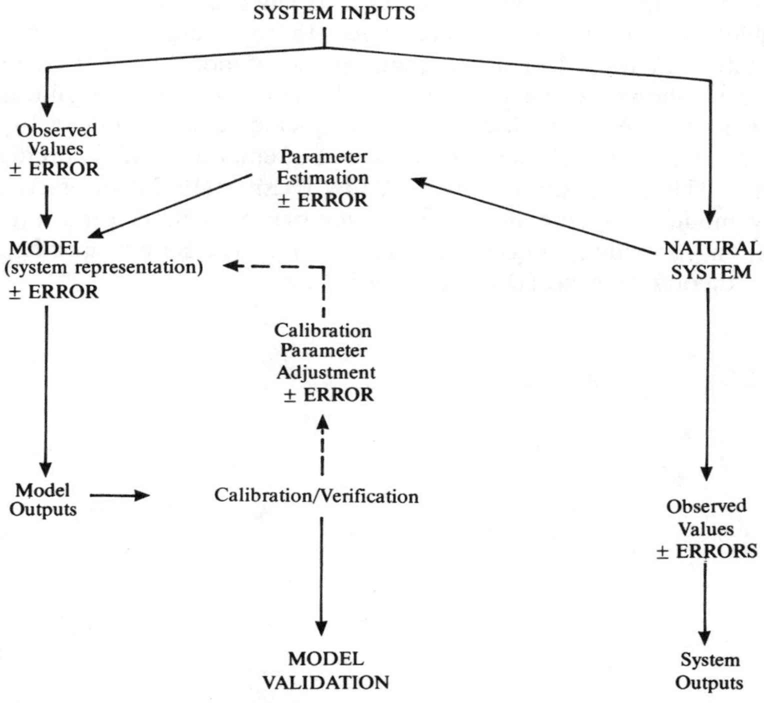

Fig. 4.11. Model and natural system inputs, outputs and errors. Reproduced from Dickson *et al.* (1982).

given task. Comparison should take into account statistical certainty expressed in the model output, input data requirements and the time and cost of code execution.

Donigian (1983) divided the process of model evaluation into a model construction check, calibration, verification and post-audit analyses. The model construction check is analogous to model and algorithm examination as defined by Shaeffer (1980). Calibration is considered to be the process of adjusting selected model parameters within an expected range until the difference between model predictions and field data are within selected criteria. Verification provides an independent test of how well the calibrated model is performing. This is achieved by a split-sample testing procedure where model predictions are compared to field observations that were not used in calibration. Finally, in post-audit, model predictions for a proposed alternative are compared to field observations following implementation of the alternatives. Acceptance criteria will reflect both the model capabilities and the assumptions made to represent the proposed alternative. Figure 4.10 illustrates the process of model evaluation and Fig. 4.11 shows how errors in model input can cause significant discrepancies between observed and predicted data. In Fig. 4.10 a distinction is made between the evaluation of empirical and theoretical models. This distinction was discussed by Lassiter (1982). In practice, many models combine theory with empiricism, with empirical formulations providing process descriptions for interactions lacking a sound theoretical basis (Donigian, 1983).

Chapter 5

The Soil Environment

5.1 INTRODUCTION

For prediction of the environmental fate of chemicals, the environment may be conceptualized as three distinct components between which, and within which, pollutants move. These three components are the terrestrial, the aquatic and the atmospheric environments. For most practical purposes, the terrestrial component is considered to be a soil environment. Contaminants spilled on the land surface will eventually be transported by physical and chemical processes to subsurface soil, or aquatic or atmospheric sinks.

Bonazountas (1983) suggested that the soil compartment is traditionally divided into three distinct subcompartments in chemical fate modelling studies. These are as follows:

A. Land surface (watershed).
B. The unsaturated soil zone (soil).
C. The saturated zone of a region (groundwater).

The soil subcompartments are illustrated schematically in Fig. 5.1. Generally, mathematical simulation is centred round two major environmental cycles, namely the hydrologic cycle and the pollutant cycle. A third cycle, sedimentation, is accounted for by land surface models. Such models describe chemical fate in the watershed, and the relative chemical distribution to a water body and to the unsaturated soil zone of the compartment. Unsaturated soil zone models describe flow and quality conditions in a soil zone profile extending between the land surface and the groundwater table. Soil zone models can be extremely complex, mainly because physical and chemical behaviour

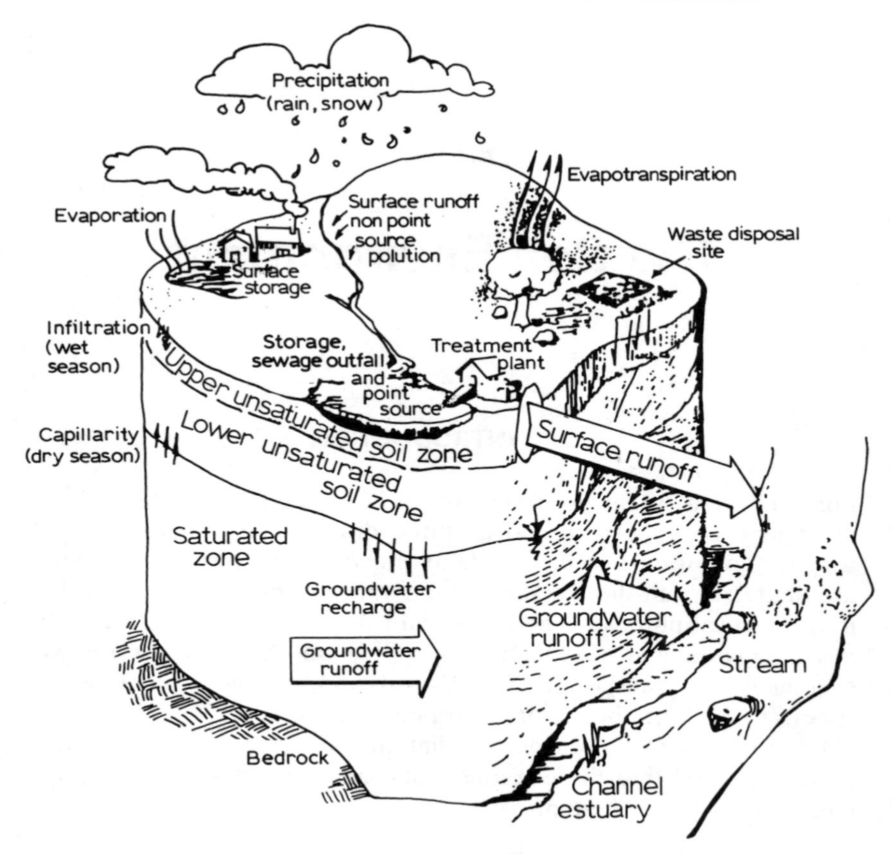

Fig. 5.1. Schematic representation of the soil compartment. Reproduced from Bonazountas and Wagner (1981).

in this subcompartment is temporally dependent on extracompartmental factors such as precipitation, air temperature and insolation.

Saturated zone models describe chemical fate in aquifers. Organic contaminants can reach the groundwater zone either dissolved in water or as organic liquid phases that may be immiscible with water. Dissolved contaminants can result from spills or leaks of aqueous solutions or from the leaching of solid phases or immiscible organic liquids present in the unsaturated (vadose) zone or land disposal areas. The subsurface transport of immiscible organic liquids is governed by a set of factors different from those for dissolved contaminants. The solubility of organic compounds varies from the infinitely miscible polar compounds, such as methanol, to the virtually insoluble nonpolar compounds such as PAHs. Halogenated aliphatics with one or two carbon atoms are typical of many groundwater contaminants, and have moderately low solubilities of less than 1% (Mackay *et al.*, 1985).

The migration of an immiscible organic liquid phase in the subsurface is governed largely by its density and viscosity. Organic liquids less dense than water tend to 'float' or spread across the water table, while those more dense than water 'sink' until reaching relatively impermeable strata (Schwille, 1981; Reinhard *et al.*, 1984).

Advection

In sand and gravel aquifers, advection is the dominant factor in the migration of dissolved contaminants. In advective groundwater processes, solutes are transported by the bulk motion of groundwater flowing from regions of the subsurface where water levels are high to regions where the water level is low (Mackay *et al.*, 1985). The magnitude of the driving force is termed the hydraulic gradient, and the average linear velocity at which groundwater flows through a granular medium is equal to the product of the gradient and the inherent capability of the medium to transmit water (Freeze & Cherry, 1979). Groundwater velocities in uniform sand and gravel aquifers under natural gradient conditions are typically between 10 and 100 m $year^{-1}$, with a potential range of between 1 and 1000 m $year^{-1}$. In the zone of influence of a high-capacity well, the average linear velocity is increased due to the artificially increased gradient. Dense organic liquids may migrate downwards with a dominant vertical component, even in horizontally flowing aquifers. Groundwater contaminant transport modelling has been reviewed by Pinder (1984) and used on field data by Barcelona and Naymik (1984) and Goerlitz *et al.* (1985).

Nonadvective processes

The transport and transformation of substances in soil and groundwater are determined by physical, chemical and biological processes associated with the substance itself and the environment in which it is found. Although numerous parameters influence such processes, modelling studies generally attempt to restrict mathematical descriptions to one or two variables (Bonazountas & Fiksel, 1982). Current practice is to model the soil compartment as though processes are subject to first-order kinetics. Thus, simulations of environmental fate frequently fail to describe adequately the environmental situation.

Of the physical processes, of paramount importance in soil modelling studies are the effects of adsorption, diffusion and volatilization. Dissolved contaminants spread as they move through groundwater. This process of dispersion results from the separate processes of molecular diffusion and mechanical mixing and is discussed later in this chapter. Of the chemical processes influencing the environmental fate of contaminants, ionization, hydrolysis, oxidation–reduction and complexation are of great significance. Biological transformation processes are considered later in Chapter 6, with respect to the aquatic environment. Details of soil compartment models are given at the end of this chapter.

5.2 PHYSICAL PROCESSES

The physical properties of a chemical often determine its environmental partitioning. Volatile chemicals will migrate to the atmospheric phase, while nonvolatile substances are likely to become sorbed to the organic fraction of the soil. Bonazountas (1983) cited the physical processes of advection, diffusion, sorption and volatilization as being important to simulations of both small, chronic chemical releases and large spills. However, bulk properties such as viscosity and solubility are believed significant only in simulations of large-scale releases.

5.2.1 Adsorption

Being the adhesion or binding of chemical ions and molecules to the surface of soil particles, adsorption processes result in an increase in chemical concentrations on soil particle surfaces with respect to chemical concentrations in interstitial waters. The significance of adsorption (sorption) and its natural converse, desorption, lies in their effect on the migration of substances through soil compartments.

Single layer adsorption is simulated by the Langmuir isotherm model. This assumes that maximum adsorption corresponds to a saturated monolayer of solute molecules on the adsorbent surface, that there is no migration of adsorbate on the surface phase, and that the energy of adsorption is constant. The Langmuir model is described by the equation (Bonazountas, 1983):

$$ds/dt = K_{sw}(s_e - s)$$

and

$$s_e = Q^{\circ}bc/(1 + c)$$

where ds/dt is the temporal variation of adsorbed concentration of compound on soil particles, s is the adsorbed concentration of compound on soil particles, K_{sw} is the Langmuir equilibrium soil–water adsorption kinetic coefficient, S_e is the maximum soil adsorption capacity, Q° is the number of moles (or mass) of solute adsorbed per unit weight of adsorbent (soil) during maximum saturation of soil, b is the adsorption partition coefficient, t is time and c is the concentration of compound in soil mixture.

The empirical Freundlich sorptive isotherm model is expressed as

$$s = x/m = Kc^{1/n}$$

where x is the adsorbed substance mass on soil, m is the mass of soil, K is the adsorption (partitioning) coefficient, c is the dissolved concentration of substance in soil moisture and n is the Freundlich equation parameter.

At very low concentrations, and particularly with organic compounds, adsorbed concentrations are presumed to be proportional to dissolved concentrations, so $1/n$ is replaced by unity. With organics, K may be replaced by the adsorption coefficient on organic carbon (K_{oc}) where

$$K = K_{oc}/100$$

In spite of being first introduced some 70 years ago, the Langmuir and Freundlich isotherms still remain the two most commonly used adsorption isotherm equations. Both can be transformed to a linear form, and so their adjustable parameters are easily estimated either by graphical means or by linear regression. Where nonlinear regression is called for, the usual procedure is to minimize the residual sums of squares, termed non-linear least squares (NLLS), or nonlinear regression. This involves finding the set of parameters that minimizes the weighted residual sum of squares (WRSS) which for m observations is given by:

$$\mathrm{WRSS} = \left(\sum_{i=1}^{m}\right) w_i(n_i - \hat{n}_i)^2$$

Where $\hat{n}_i$ is the fitted value for observation i calculated from a particular isotherm equation and w_i is the associated weighting factor (Kinniburgh, 1986). Some useful adsorption isotherm equations are summarized in Table 5.1.

Table 5.1
Some useful adsorption isotherm equations[a]

Isotherm	*Equation*	*Number of adjustable parameters*	*Asymptotic properties*	
			Linear at low c	*Adsorption max. at high c*
Langmuir	$n = KcM/(1+Kc)$	2	Yes	Yes = M
Freundlich	$n = (Kc)^{\beta}$	2	No	No
Langmuir–Freundlich	$n = (Kc)^{\beta}M/[1+(Kc)^{\beta}]$	3	No	Yes = M
Redlich–Peterson	$n = KcM/[1+(Kc)^{\beta}]$	3	Yes	No
Toth	$n = KcM/[[1+(Kc)^{\beta}]1/\beta]$	3	Yes	Yes = M
Multisite Langmuir	$n = \sum_{i=1}^{k} [K_i c M_i/(1+K_i c)]$	$2k$	Yes	Yes = $\sum M_i$
Sum of Freundlich's	$n = \sum_{i=1}^{k} (K_i c)^{\beta i}$	$2k$	No	No
Dubinin–Radushkevich	$\log n = -\beta[\log^2(Kc)] + \log M$	3	No	No
Modified Dubinin–Radushkevich[b]	$\log n = \begin{cases} \log n^* - \log c^* + \log c & c < c^* \\ -\beta \log^2[Kc/(1+Kc)] + \log M & c > c^* \end{cases}$	3	Yes	Yes = M

[a] n is the amount adsorbed, c is the equilibrium solution concentration, M is the adsorption maximum, k is an affinity parameter, and β is an empirical parameter that varies with the degree of heterogeneity.
[b] Where (n^*, c^*) is the point where $\partial \log n/3 \log c = 1$; c^* cannot be calculated explicitly, but is given to a good approximation by $\log c^* \approx (-1/2\beta) - \log K$. This estimate can be refined by a modified Newton using $\log c^*_{i+1} = \log c^*_i + (b_i - 1)/2\beta$ where $b_i = [-2\beta/(1+Kc^*_i)]\log[Kc^*_i/(1+Kc^*_i)]$ and i is the iteration number.
Reproduced from Kinniburgh (1986).

Karickhoff *et al.* (1979) showed that differences in the sorption of hydrophobic compounds, such as the aromatic and chlorinated hydrocarbons, within silt and clay fractions in sediments, were largely due to differences in organic carbon content. Organic carbon adsorption coefficients (K_{oc}) could be estimated from octanol–water partition coefficients (K_{ow}) and the fractional mass of organic carbon in the sediment (OC) as:

$$K_{oc} = K_{ow}/OC$$

Since K_{oc} can be determined independently of other soil properties, some prediction of the movement of a chemical through any soil is possible if the organic carbon content is known. The influence of the organic carbon content of the sorbent has been discussed with respect to chlorinated phenols (Schellenberg *et al.*, 1984), anthracene and two herbicides (Nkedi-Kizza *et al.*, 1985) and hexachlorobiphenyl (Di Toro *et al.*, 1985). Hamaker (1975) demonstrated that chromatographic theory can be used to correlate sorption coefficients with leaching potential in soil column experiments by means of the following derived equation:

$$R = 1/\theta^{2/3}[1 - K_{oc}(\%OC/100)d_s] + K_{oc}(\%OC/100)d_s$$

where R is the ratio of centimetres moved by the chemical to centimetres of water entering soil, θ is the pore fraction of the soil, d_s represents bulk density of the soil solids, K_{oc} is the sorption coefficient (g chemical/g organic carbon/g chemical per gram of water) and $\%OC$ is the per cent organic carbon.

If $K_{oc}(\%OC/100)d_s$ is significantly greater than unity, then

$$R = 1/K_{oc}(\%OC/100)(1 - \theta^{2/3})d_s$$

or

$$R = 1/K_d(1 - \theta^{2/3})d_s$$

where K_d equals $K_{oc}(\%OC/100)$ with units of μg chemical per gram of soil/μg chemical per gram of water. Rearranging, it follows that:

$$\text{centimetres moved by chemical} = \frac{1}{K_d}\left(\frac{\text{cm } H_2O \text{ entering soil}}{(1 - \theta^{2/3})d_s}\right)$$

Thus, for a given soil with a specific amount of water a plot of $1/K_d$ versus centimetres moved should be linear for a number of chemicals.

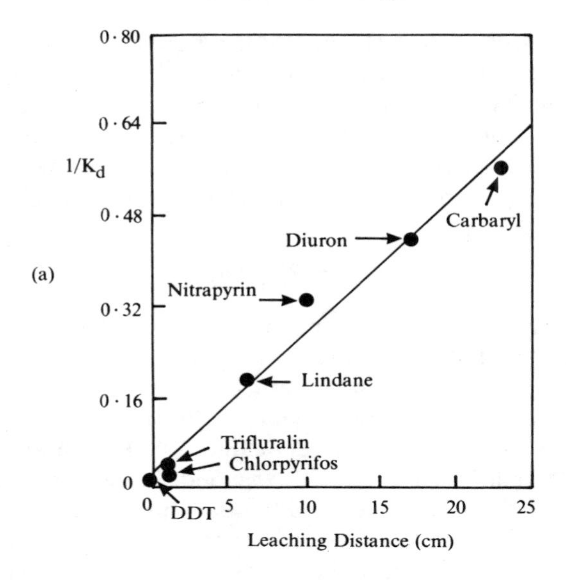

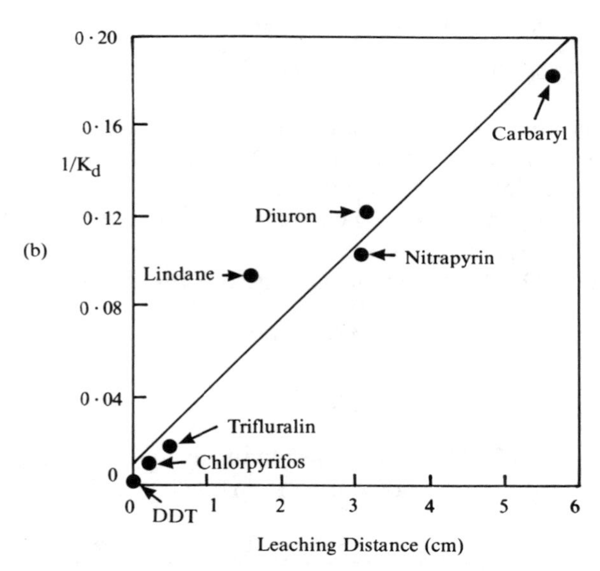

Fig. 5.2. Plots of $1/K_a$ versus leaching distances for seven chemicals in (a) commerce and (b) catlin soil. Reproduced from McCall *et al.* (1981).

Details of just such a study using a variety of chemicals leached through columns containing various soils have been given by McCall *et al.* (1981). Properties of two agricultural soils 'Commerce' and 'Catlin' were, respectively, organic carbon 0·68%, 2·01%; pH 6·7, 6·2; sand 38%, 12%; silt 48%, 56%; clay 14%, 32%. It can be seen from Fig. 5.2 that leaching distance was much reduced in the silty Catlin soil with its higher percentage of organic carbon content. None the less, the relative positions of the seven chemicals on the linear plot remained unchanged. In a conceptually similar experimental study, a good correlation ($r = 0{\cdot}96$) was obtained when the natural logarithm of K_{oc} was plotted against the natural logarithm of a reverse-phase high-performance liquid chromatographic (HPLC) retention time (Swann *et al.,* 1981). Such a correlation was deemed adequate argument for the use of HPLC in the estimation of K_{oc} values. The sorption and leaching of the two herbicides diuron and atrazine has been measured in soil columns eluted with aqueous solutions and binary solvent mixtures of methanol and water (Nkedi-Kizza *et al.,* 1987).

A more theoretical approach to determining soil sorption coefficients is to utilize prediction based on molecular topology. Tests on some 190 polar and ionic compounds, including PAHs, alkylbenzenes, chlorobenzenes and halogenated phenols, have shown that a molecular connectivity model plus a single semiempirical variable is capable of accounting for the soil sorption properties of nearly 95% of all organic chemicals whose soil sorption coefficients have been measured (Sabljic, 1987). The sorption of linear alkylbenzenesulphonates, used primarily as surfactants in detergents and cleaning products, to river sediment increased predictably with chain length and phenyl position (Hand & Williams, 1987). Finally, studies on the effects of deoxyribonucleic acid (DNA) molecular size on adsorption to sand and a sandy soil were described by the Freundlich isotherm model ($r^2 > 0{\cdot}85$) and indicated a direct relationship between molecular weight and adsorption (Ogram *et al.,* 1988).

Ion exchange

An important sorption mechanism for inorganic chemical groups is ion exchange. This may be seen as an exchange process with some other ion that initially occupies the adsorption site on the soil particle. For divalent cations (M^{2+}) in clay, the exchange ion is often calcium, where:

$$M^{2+} + [\text{clay}].Ca \rightleftharpoons Ca^{2+} + [\text{clay}].M$$

Bonazountas (1983) and Bonazountas and Wagner (1981) suggested that the maximum mass of a substance associated with the solid phase (S) is given by the equation:

$$S = EC.MW/VAL$$

where S has units of mass chemical/mass of solid, EC is the cation exchange capacity (mass equivalents/mass dry soil), MW is the molecular (or atomic) weight of the chemical (mass/mole) and VAL represents the valence of the ion.

Details of competitive adsorption during solute transport in soils have been given in a review of the experimental evidence by Murali and Aylmore (1983). Although competition was influenced by factors such as pH, solution concentration and the nature of the competing species, a number of competitive adsorption models satisfactorily predicted data from experimental isotherms. Reproduced from the review of Rao and Jessup (1982) on pesticide dynamics simulation models, Table 5.2 presents a summary of retention and transformation models for soils.

5.2.2 Diffusion

Molecular diffusion is the net transport of a molecule in a liquid or gas medium and is the result of intermolecular collisions rather than turbulence or bulk transport. The process is driven by gradients such as pressure, temperature and concentration. The rate of diffusion depends on both the nature of the compound and the medium through which it moves. The diffusion coefficient, or diffusivity, was defined by Tucker and Nelken (1982) as

$$\mathscr{D}_{BA} = J_B/\nabla X_B$$

where $\mathscr{D}_{BA}$ is the diffusion coefficient of compound B in compound A ($cm^2\,s^{-1}$), J_B is the net molal flux of B across a hypothetical plane ($mol\,cm^{-2}\,s^{-1}$) and ∇X_B is the concentration gradient of B at the hypothetical plane ($mol\,cm^{-3}\,cm^{-1}$). The cursive $\mathscr{D}$ is used to distinguish this property from the *apparent* diffusion coefficient (D).

In the atmosphere or in surface water bodies, dispersion is controlled by the intensity of turbulent mixing rather than molecular diffusion and under these conditions D is the appropriate parameter.

Table 5.2
Summary of models for pesticide retention and transformations

	Equation
Equilibrium models	
1.1 (linear)	$S = k_1 C + k_2; k_2 \geq 0$
1.2 (Langmuir)	$S = k_1 C/(1 + k_2 C)$
1.3 (Freundlich)	$S = k_1 C^{k_2}$
1.4	$S = k_1 C e^{-2k_2 s}$
1.5 (Modified Kjelland)	$S = CS_m/\{C + k_1(C_m - C)\exp[k_2(C_m - 2C)]\}$
1.6 (Fraction near equilibrium)	$S = (\text{FREQ})k_1 C^{k_2}$
Kinetic models	
2.1 (linear)	$\partial S/\partial t = k_r(k_1 C + k_2 - S)$
2.2 (Langmuir)	$\partial S/\partial t = k_r[(k_1 C)/(1 + k_2 C) - S]$
2.3 (Freundlich)	$\partial S/\partial t = k_r(k_1 C^{k_2} - S)$
2.4	$\partial S/\partial t = k_r e^{k_2 s}(k_1 C e^{-2k_2 s} - S)$
2.5	$\partial S/\partial t = k_r(S_m - S)\sin h[k_2(S_m - S)/(S_m - S_i)]$
2.6 (Two-site)	$\partial S/\partial t = \partial S_1/\partial t + \partial S_2/\partial t$, where
	$\partial S_1/\partial t = k_1 C^{k_2}\, \partial C/\partial t$
	$\partial S_2/\partial t = k_r(k_1 C^{k_2} - S_2)$
Degradation models	
3.1 (First-order)	$Q = k_d(\Theta C + \rho S)$
3.2 (Power rate)	$Q = k_{d_1} C^k d_2$
3.3 (Hyperbolic)	$Q = k_{d_1}/(^k d_2 + C)$

C, Solution-phase concentration (ML^{-3}); C_m, maximum solution concentration (ML^{-3}); D, pesticide dispersion coefficient (L^2T^{-1}); FREQ, fraction of adsorption sites in equilibrium; h, soil-water potential (L); $K(\Theta)$, soil hydraulic conductivity (LT^{-1}); k_1, k_2, k_3, empirical adsorption isotherm constants; k_{d_1}, degradation rate constants (T^{-1}); k_{d_2}, empirical degradation constant; k_r, adsorption-desorption rate constant (T^{-1}); q, Darcy soil water flux (LT^{-1}); Q, sink term for pesticide degradation ($ML^{-2}T^{-1}$); S, adsorbed-phase concentration (MM^{-1}); S_1, adsorbed concentration on 'fast' sites (MM^{-1}); S_2, adsorbed concentration on 'slow' sites (MM^{-1}); S_i, initial adsorbed concentration (MM^{-1}); S_m, maximum adsorbed concentration (MM^{-1}); t, time (T); $t_{1/2}$, half-life for pesticides in soil (T); U, sink term for soil-water uptake by roots (T^{-1}); v, average pore-water velocity (LT^{-1}); z, distance (L); Θ, volumetric soil-water content (L^3L^{-3}); ρ, soil bulk density (ML^{-3}).
Reproduced from Rao and Jessup (1982).

However, in certain special environmental situations, such as air–water interfaces, the interstitial waters of sediments and groundwater (saturated or unsaturated soils), molecular diffusion is a significant factor in determining chemical fluxes. In a review of exchange at the sediment–water interface, Lerman (1978) discussed data of Manheim (1970), which indicate that the ratio of the apparent diffusion coefficient to the molecular diffusion coefficient varies as the sediment porosity is raised to the nth power, where $1{\cdot}2 < n < 2{\cdot}8$.

In saturated ground waters (aquifers) the effective dispersivity is generally much greater than the molecular diffusion coefficient (Tucker & Nelken, 1982). Scheidigger (1974) postulated that the apparent dispersivity is proportional to the flow velocity and that molecular diffusion may therefore be significant in aquifer systems with very slow flow rates. Tucker and Nelken (1982) considered that molecular diffusion could be ignored at pure water velocities exceeding $0{\cdot}002\ \mathrm{cm\,s^{-1}}$. Harleman and Rumer (1962), cited in Kirda *et al.* (1973), proposed the following relationship:

$$D_s = f_t \mathscr{D}_{BW} + \alpha u^m$$

where D_s is the apparent dispersion coefficient ($\mathrm{cm^2\,s^{-1}}$), f_t is the tortuosity factor—a soil property (0·01–0·50), u is the pore water velocity ($\mathrm{cm^2\,s^{-1}}$), α and m are empirically determined soil constants and BW represents compound B in water.

Unsaturated soils demand a more complicated analysis, since these represent a triphasic system consisting of soil particles, water and air. Thus, dispersion is determined by liquid- and gas-phase diffusion as well as diffusion along the water–air and water–solid interfaces. Shearer *et al.* (1973) rigorously equated the effective diffusion coefficient (D_{EF}) with:

$$D_{EF} = \left(\frac{\mathscr{D}_{AB} P^{7/3}}{P_T^2 (S+1)}\right) + \left(\frac{S}{1+S}\right)\left(\frac{D_s + D_H K'\beta + \beta D_I S'}{\beta K' + \theta + \beta S'}\right)$$

where $\mathscr{D}_{AB}$ is the gas diffusion coefficient in air ($\mathrm{cm^2\,s^{-1}}$), P is the air-filled porosity of the soil (0–$0{\cdot}7\ \mathrm{cm^3\,cm^{-3}}$), P_T is the total porosity of the soil (0–$0{\cdot}7\ \mathrm{cm^3\,cm^{-3}}$), S is the equilibrium coefficient of proportionality between vapour mass density and total concentration of (pesticide) in soil ($\mathrm{g\,cm^{-3}}$)/($\mathrm{g\,cm^{-3}}$), D_s is the apparent solution phase diffusion coefficient ($\mathrm{cm^2\,s^{-1}}$), D_H is the apparent diffusion coefficient of molecules adsorbed at the solution–solid interface ($\mathrm{cm^2\,s^{-1}}$), K' is the sorption coefficient ($\mathrm{cm^3\,g^{-1}}$), β is the soil bulk density ($\mathrm{g\,cm^{-3}}$), D_I is the apparent diffusion coefficient of molecules adsorbed at the air–solution interface ($\mathrm{cm^2\,s^{-1}}$), S' is the coefficient of proportionality between solution concentration and vapour concentration at the air–solution interface ($\mathrm{cm^3\,s^{-1}}$) and θ is the fractional volumetric water content ($\mathrm{cm^3\,cm^{-3}}$).

Furthermore, Scheibel (1954) showed that

$$D_s = (\theta / P_T)^2 \theta^{4/3} \mathscr{D}_{BW}$$

Where liquid-phase diffusion is assumed, Walker and Crawford (1970) developed the equation:

$$D_{\mathrm{EF}} = \frac{\mathscr{D}_{\mathrm{BW}}\theta f_t}{\beta K' + \theta}$$

In this case, where $\mathscr{D}_{\mathrm{BA}} = D_{\mathrm{H}} = S' = 0$

$$D_{\mathrm{EF}} = \frac{S[(\theta/P_{\mathrm{T}})^2\theta^{4/3}\mathscr{D}_{\mathrm{BW}}}{(1+S)(\beta K' + \theta)}$$

Thus, the effective diffusion coefficient is proportional to the molecular diffusion coefficient and inversely proportional to the term $(\beta K' + \theta)$. Six methods for estimating the liquid diffusion coefficient have been summarized in Table 5.3.

5.2.3 Volatilization

Volatilization is the process whereby a compound evaporates in the vapour phase to the atmosphere from another component of the environment. Volatilization is an extremely important form of diffusion for many organic chemicals, the process being less significant for inorganic substances. Most ionic compounds are considered nonvolatile (Bonazountas, 1983).

Three categories of factor affect the volatilization process (Spencer *et al.* 1973), namely

A. Those which affect movement away from the evaporating surface into the atmosphere.
B. Those which affect the vapour density of the chemical.
C. Those which control the rate of movement to the evaporating surface.

The partitioning of a chemical between the three environmental phases, air, soil and water, which together comprise the soil system, can be estimated from either vapour- or solution-phase desorption isotherms (Spencer, 1970). Adsorption may affect the vapour density and the volatilization rate (Spencer & Cliath, 1970). Guenzi and Beard (1974) showed that the vapour density is inversely related to both the surface area of the soil particles and the organic matter content of the soil. The water content of a soil may also affect volatilization losses by competing for soil sorption sites (Spencer, 1970; Guenzi & Beard, 1974). Temperature primarily affects vapour density, with an increase in temperature normally increasing the equilibrium vapour density

Table 5.3
Available methods for estimating diffusivity into water

Method	*Formula*	*Inputs (excluding water parameters)*	*Absolute average error (%) (Hayduk & Laudie, 1974)*
Hayduk and Laudie (1974)[a]	$\partial_{BW} = \dfrac{13{\cdot}26 \times 10^{-5}}{n_W^{1{\cdot}14} V_B'^{0{\cdot}589}}$	V_B	5·8 (87 solutes)
Wilke and Chang (1955)	$\partial_{BW} = \dfrac{7{\cdot}4 \times 10^{-8}(\phi_W M_W)^{1/2} T}{n_W V_B^{0{\cdot}6}}$	V_B	8·8 (87 solutes)
Scheibel (1954)	$\partial_{BW} = \dfrac{8{\cdot}2 \times 10^{-8} T}{n_W V_B^{1/3}} \left[1 + \left(\dfrac{3V_W}{V_B}\right)^{2/3}\right]$	V_B	6·7 (87 solutes)
Othmer and Thakar (1953)	$\partial_{BW} = \dfrac{1{\cdot}4 \times 10^{-5}}{n_W^{1{\cdot}1} V_B^{0{\cdot}6}}$	V_B	5·9 (87 solutes)
Reddy and Doraiswamy In: Reid *et al.* (1977)[b]	$\partial_{BW} = \dfrac{M_W^{1/2} T K'}{n_W (V_W V_B)^{1/3}}$	V_B	20 (96 solutes)
Venezian (1976)	$\partial_{BW} = \dfrac{6 \times 10^{-10} T}{n_W (R_M - 0{\cdot}855)}$ where $R_M = \left(\dfrac{n_D^2 - 1}{n_D^2 + 2}\right)\left(\dfrac{M_B}{\rho_B}\right)^{1/3}$	n_D, ρ_B M_B	

[a] Recommended by Tucker and Nelken (1982).
[b] $K' = 10 \times 10^{-8}$ for $V_W/V_B \leqslant 1{\cdot}5$ and $8{\cdot}5 \times 10^{-8}$ for $V_W/V_B > 1{\cdot}5$.
Data from Tucker and Nelken (1982).

(Thomas, 1982*a*). Atmospheric conditions determining air flow over the soil surface play an important role in the removal of vapour from the soil surface (Hartley, 1969).

Organic compounds can undergo a variety of equilibrium and transport processes in the subsurface zone (Jury *et al.*, 1983; Mackay *et al.*, 1985; Bauer, 1987). Dissolved volatile organic compounds (VOCs) tend to partition into the soil atmosphere because of their relatively low solubilities and high vapour pressure. Thus, soil–gas investigations can be used to identify sources of VOCs, and to distinguish between soil and groundwater contamination (Marrin, 1988). Solvent and fuel leaks from underground storage tanks can also be monitored by chemical analysis of soil gases.

The solution/vapour equilibrium of VOCs is quantitatively described by Henry's Law. Dividing the vapour pressure of a pure compound by its water solubility gives the Henry's Law constant. This, when multiplied by the dissolved concentration in a solution, provides an estimate of the VOC concentration in the adjacent gas phase. Organic compounds with the highest Henry's Law constants will partition most favourably from groundwater to the soil gases. VOC equilibrium across the capillary fringe is also affected by solute and matrix properties, so Henry's Law constants provide only a relative indication of solution/vapour partitioning in the field (Marrin & Kerfoot, 1988). Once VOCs enter the soil gas, they diffuse in response to a chemical concentration gradient. VOC concentrations in soil gas above any diffusion limiting layer, such as saturated clays, perched water and pavement, will be lower than concentrations below such a layer.

The contribution of municipal landfills and incinerators to PCB emissions to the atmosphere in the Great Lakes region has been estimated by Murphy *et al.* (1985). A rate of 10 to 100 kg year^{-1} from sanitary landfills was small compared to the 900 000 kg year^{-1} of PCBs estimated to cycle through the atmosphere over the USA annually. Table 5.4 lists the most frequently encountered substances at Superfund sites and their Henry's Law constants. Volatilization from water is considered in Chapter 6, which also provides a fuller account of the role of Henry's Law in volatilization studies.

Table 5.5 and the Appendix to this chapter summarize the main features and requirements of models used to calculate the volatilization of chemicals from soil. Reproduced from Thomas (1982*a*), Fig. 5.3 and the Appendix provide a guide to the selection of an

Table 5.4
Most frequently identified substances at 546 'Superfund' hazardous waste sites in the USA

	Henry's law constant[a]		Percentage of sites
	ppbv . litres µg[b]	*Dimensionless*[c]	
Trichloroethylene[d]	72	0·385	33
Lead and lead compounds	NA	NA	30
Toluene[d]	56	0·230	28
Benzene[d]	71	0·245	26
Polychlorinated biphenyls (PCBs)	≪1	≃0	22
Chloroform[d]	40	0·111	20
Tetrachloroethylene[d]	123	0·588	16
Phenol	≪1	≃0	15
Arsenic and arsenic compounds	NA	NA	15
Cadmium and cadmium compounds	NA	NA	15
Chromium and chromium compounds	NA	NA	15
1,1,1-Trichloroethane[d]	30	0·667	14
Zinc and zinc compounds	NA	NA	14
Ethylbenzene[d]	59	0·279	13
Xylenes[d]	32	0·203	13
Methylene chloride[d]	23	0·083	12
trans-1,2-Dichloroethylene[d]	77	0·333	11

[a] NA = not applicable.
[b] Gas-phase concentration (ppbv) ÷ liquid-phase concentration (μg litre^{-1}).
[c] Gas concentration (μg . litre^{-1}) ÷ liquid concentration (μg litre^{-1}).
[d] Amenable to soil-gas sampling.
Data from Kerfoot and Barrows (1986).

appropriate volatilization model. A list of volatile organic chemicals to be found in both indoor and outdoor air may be found in a review by Shah and Singh (1988).

5.3 CHEMICAL PROCESSES

5.3.1 Ionization

An acid or base that is extensively ionized may have markedly different solubility, sorption, toxicity and biological properties from the corresponding neutral compound. For example, the ionized species

Table 5.5
Models used to compute volatilization of chemicals from soil

Method	*Equations*	*Difficulty (calculational)*	*Information required*	
			Chemical	*Environmental*
Hartley (1969)	1, 2	Low	Saturated vapour concentration Vapour diffusion coefficient in air Latent heat of vaporization Molecular weight of chemical Thermal conductivity of air Gas constant	Humidity Stagnant air layer thickness Temperature
Hamaker (1972*b*)			Vapour diffusion coefficient in soil	Initial concentration
No water loss	3	Low	Vapour pressure of chemical Vapour pressure of water	Water flux from plot—both liquid and vapour phase
Water loss	4	Low	Vapour diffusion coefficient in soil—for chemical and water	
Mayer *et al.* (1974) (five models)	5	High	Air/soil concentration isotherm coefficient Diffusion coefficient in soil Diffusion coefficient in air	Depth of soil column Air flow velocity Initial concentration Adsorbed concentration Thickness of nonmoving air layer
Jury *et al.* (1980)	6, 7, 8	Medium	Gas-phase diffusion coefficient Liquid-phase diffusion coefficient Adsorption parameters Henry's law constant	Total concentration in soil Soil bulk density Adsorbed concentration Volumetric soil water content Chemical concentration in liquid phase Chemical concentration in gas phase Soil air content Water flux from plot
DOW (Swann *et al.*, 1979)	9, 10, 11	Low	Soil adsorption coefficient Vapour pressure Solubility	

Data from Thomas (1982*a*).

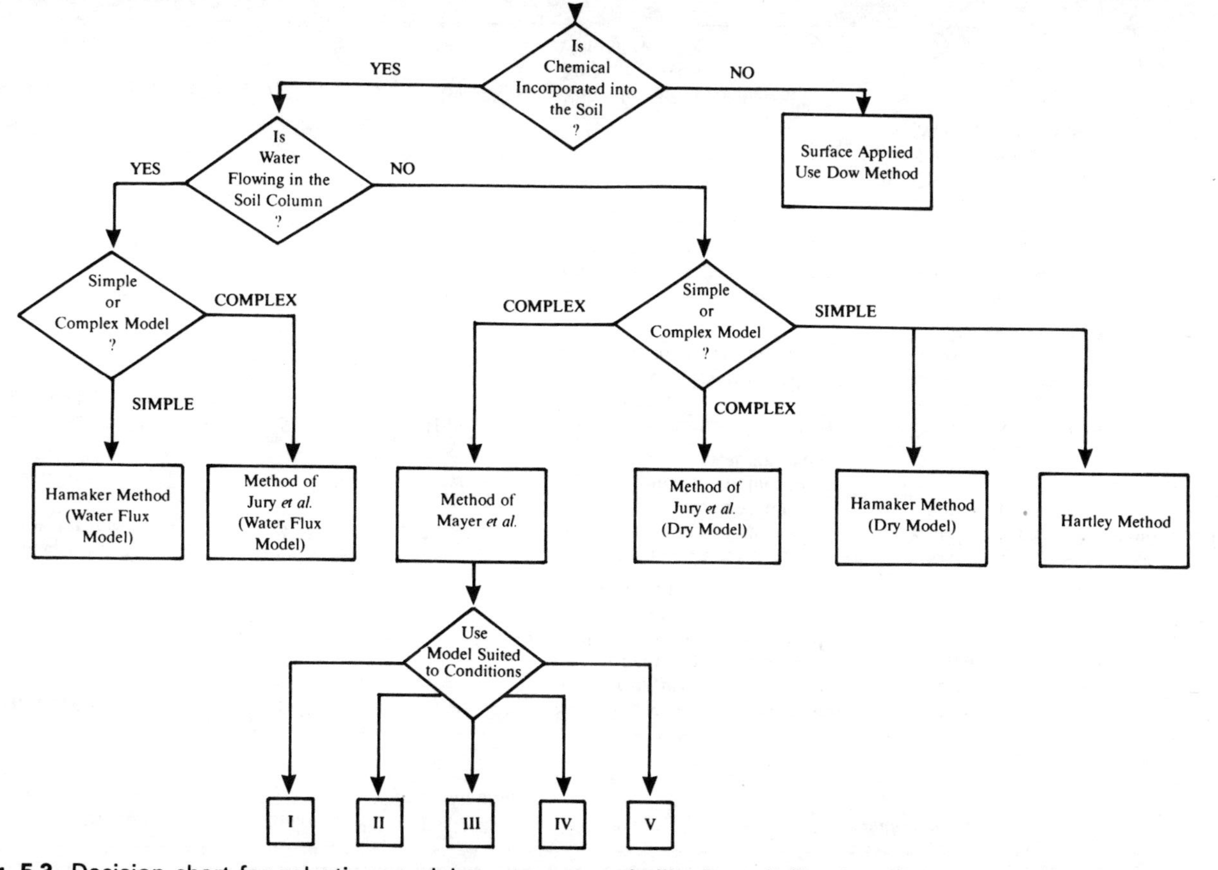

Fig. 5.3. Decision chart for selecting model to compute volatilization of chemicals from soil. Reproduced from Thomas (1982*a*).

of an organic acid is generally adsorbed by sediments to a much lesser degree than is the neutral form (Harris & Hayes, 1982). A weak acid HA will ionize in water according to the reaction:

$$HA + H_2O \rightleftharpoons H_3O^+ + A^-$$

The acid dissociation constant K_a is defined as the equilibrium constant for this reaction, thus

$$K_a = a_{H_3O^+} a_{A^-} / a_{HA} a_{H_2O}$$

where a_i is the activity of any species i in an aqueous solution.

Conventionally, unit activity is assigned to solvent water, thereby designating pure water as the standard state and assuming that the solution is so dilute that the presence of solute has no effect on the activity of water. Such an assumption is generally true for solute concentrations less than 0·1 M. Then K_a reduces to

$$K_a = a_{H_3O} + a_{A^-} / a_{HA} = (\gamma_{H_3O} + M_{H_3O^+})(\gamma_{A^-} M_{A^-}) / \gamma_{HA} M_{HA}$$

where γ_i is the molar activity coefficient of species i and M_i is the molar concentration of species i.

Where K_a is defined in terms of activities, it is referred to as the 'true' or thermodynamic dissociation constant. If it is assumed that all activity coefficients are unity, then K_a approximates to

$$K_a \approx M_{H+} M_{H-} / M_{HA} \qquad (\text{mol litre}^{-1})$$

Such a concentration dissociation constant provides a reasonable approximation of the thermodynamic constant for solute concentrations below 0·01 M (Harris & Hayes, 1982). A compound is 50% dissociated when the pH of the water equals the pK_a (p$K_a = -\log K_a$).

Sposito (1984) conceded that single-ion activities appear to be of increasing importance in the study of soil–plant relationships and in the surface chemistry of soils. This was in spite of the fact that the chemical potential of a charged species cannot be defined in a strict thermodynamic sense, and that no general procedure exists through which the activity of a charged species can be related to macroscopic properties without using unmeasurable parameters. Sposito (1984) went on to outline a general theoretical constraint on single-ion activity coefficients, incorporating a discussion of Young's rules (Whitfield, 1979), the principle of specific interaction (Guggenheim, 1967), the Pitzer equation (Pitzer, 1977, 1979) and the Davies equation (Davies, 1962).

5.3.2 Hydrolysis

Under environmental conditions, hydrolysis occurs mainly with organic compounds (Bonazountas, 1983). Hydrolysis is a chemical transformation process in which an organic molecule, RX, reacts with water forming a new carbon–oxygen bond and cleaving a carbon–X bond in the original molecule (Harris, 1982*a*). The net reaction is most commonly direct displacement of X by OH:

$$\mathrm{R{-}X} \xrightarrow{\mathrm{H_2O}} \mathrm{R{-}OH} + \mathrm{X^-} + \mathrm{H^+}$$

Hydrolysis should be distinguished from acid:base reactions, carbonyl hydration (reversible), addition to carbon–carbon bonds (reaction conditions unlikely in the environment) and elimination reactions, which are generally favoured by higher temperatures and more strongly basic conditions than are typical of aqueous environments. Elimination may compete with hydrolysis for organic compounds containing good leaving groups such as halides. Hydrolysis comprises a family of reactions involving such diverse compound types as alkyl halides, carboxylic acid esters, carbonates and nitrites. Table 5.6 lists some types of organic functional groups that are generally resistant to hydrolysis and some that are potentially susceptible.

Hydrolysis reactions may be modelled as first-order processes using pseudo first-order rate constants $k_T(\mathrm{s}^{-1})$ where

$$-\mathrm{d}[\mathrm{RX}]/\mathrm{d}t = k_T[\mathrm{RX}]$$

The first-order dependence implies that the hydrolysis half-life of RX ($t_{1/2} = 0{\cdot}693/k_T$) is independent of the RX concentration.

As rates of hydrolysis of organic chemicals can vary over many orders of magnitude, with associated environmental half-lives of a few seconds to thousands of years, it is *essential* that laboratory rate constants used in soil models are corrected for environmental conditions.

5.3.3 Oxidation/reduction

Degradation oxidation reactions under environmental conditions often depend on reactions with free radicals already in solution (Bonazountas, 1983) and are usually modelled by pseudo first-order kinetics:

$$-\mathrm{d}[\mathrm{X}]/\mathrm{d}t = K_o'[\mathrm{RO_2^{\cdot}}][\mathrm{X}] = K_{ox}[\mathrm{X}]$$

where X is the pollutant, K_o' is the second-order oxidation rate

Table 5.6
Types of organic functional groups

Generally resistant to hydrolysis[a]	*Potentially susceptible to hydrolysis*
Alkanes	Alkyl halides
Alkenes	Amides
Alkynes	Amines
Benzenes/biphenyls	Carbamates
PAHs	Carboxylic acid esters
Heterocyclic PAHs	Epoxides
Halogenated aromatics/PCBs	Nitriles
Dieldrin/aldrin, etc.	Phosphonic acid esters
Aromatic nitro-compounds	Phosphoric acid esters
Aromatic amines	Sulphonic acid esters
Alcohols	Sulphuric acid esters
Phenols	
Glycols	
Ethers	
Aldehydes	
Ketones	
Carboxylic acids	
Sulphuric acids	

[a] Multifunctional organic compounds may be hydrolytically reactive if they contain an additional hydrolysable functional group.
Data from Harris (1982*a*).

constant, $RO_2^{\cdot}$ is a free radical and K_{ox} is a pseudo first-order oxidation rate constant.

5.3.4 Complexation

Complexation, or chelation, occurs when metal ions and organic or other nonmetallic molecules (ligands) combine to form stable metal–ligand complexes. Such complexes may prevent reactions or interactions normally associated with free metal cations. An important group of compounds acting as chelating agents in the soil compartment are the humic substances. These include the humic acids, which are soluble in basic solutions and insoluble in acids and ethanols, fulvic acid, soluble in both basic and acid solution, and humin, soluble in neither (Saar & Weber, 1982). A variety of models have been used to explain complexation of metal ions by fulvic acid. In the Scatchard method (Scatchard, 1949; Sposito, 1981*a*), fulvic acid is modelled with distinct classes of site, each with a common ability to complex metal ions.

Potentiometric studies on the neutralization of several fulvic acid sources with standard base in aqueous and nonaqueous media have ascribed the polyelectrolyte properties observed to a rather inflexible fulvic acid molecule whose variably charged surface is impermeable to simple electrolytes (Ephraim *et al.*, 1986). The influence of polyelectrolyte properties and functional heterogeneity on copper(II) ion binding equilibria in an Armadale Horizons Bh fulvic acid sample has been reported by Ephraim and Marinsky (1986). The kinetics and effect of fulvic acid on aluminium fluoride complexation in acidic waters have been reviewed by Plankey and Patterson (1987, 1988). It was found that an *a priori* model of the fulvic acid and fluoride ion ligands being in competition for aluminium was incorrect. In actual fact, the rate of fluoride ion consumption was increased by the presence of fulvic acid, probably due to mixed-ligand equilibria.

Quantitative measurements of the binding of organic compounds to dissolved humic and fulvic acids showed that the extent of binding increases as a compound's octanol–water partition coefficient increases or as its water solubility decreases (Carter & Suffet, 1983). In general, it was found that humic acids bind compounds to a greater extent than do fulvic acids. Chiou *et al.* (1986) reported that solute partition coefficient values of solutes with soil-derived humic acid were approximately four times greater than with soil fulvic acid and five to seven times greater than with aquatic humic and fulvic acids.

Partition coefficients for the binding of pyrene to 14 different humic and fulvic acids have been determined by Gauthier *et al.* (1987). Partition coefficients normalized to the fraction of organic carbon in the humic material (K_{oc} values) varied by as much as an order of magnitude depending on the structure and composition of the humic material. The magnitude of the K_{oc} correlated strongly with the degree of aromaticity of the humic acid. Caution should be exercised with experiments designed to evaluate the nature and reactivity of humic substances in natural waters and soils if commercial 'humic acids' are used, since these may not mimic exactly the behaviour of true soil and water humic substances (Malcolm & McCarthy, 1986).

Perdue and Lytle (1983) suggested that the acidic properties of humic substances are attributable primarily to carboxyl and phenolic hydroxyl groups. The latter authors went on to conclude that the complex mixture of ligands that are involved in proton binding and metal binding by aquatic humus could not be described unambiguously by any type of chemical model then available.

None the less a model was proposed in which a continuum of $\log K$ (K = conditional stability constant) values for proton or metal binding could be treated. In the model, it was assumed that ligand concentrations are normally distributed with respect to their $\log K$ values. A 'class' of ligands was then characterized by its mean $\log K$ value and the standard deviation of $\log K$ values about the mean.

The theory underlying discrete and continuous multiligand models for metal–humate binding has been reviewed by Dzombak *et al.* (1986). A comparison of models indicated that discrete ligands, typically two, are a simple and accurate means of predicting metal–humate binding within the range of experimental calibration titrations. Discrete ligand data have the added advantage of being easily incorporated into chemical equilibrium computer programs (Fish *et al.*, 1986). Correspondence relating the previous two papers has been entered into by Ruzic (1987) and Morel *et al.* (1987). The theories of complexing equilibria in mixed-ligand systems have been tested by computer simulations (Gamble & Langford, 1988).

Information from 33 studies of cadmium adsorption on a variety of substrates was collated by Laxen (1983*a,b*). The results were transferred to a unified database using a surface complexational model, and presented in terms of conditional adsorption constants as a function of pH. The affinity of the substrates, on a weight basis, for cadmium followed the order:

$$\mathrm{Mn} > \mathrm{Fe_{am}} > \mathrm{chlorite} > \mathrm{Fe_{cr}} = \mathrm{illite} = \mathrm{humics} > \mathrm{kaolinite} > \mathrm{silica}$$

where Mn is manganese oxide, Fe_{am} is amorphous hydrous iron oxide, chlorite is similar to montmorillonite (a clay mineral), and Fe_{cr} is crystalline hydrous iron oxide.

The adsorption constants for a given substrate at a fixed pH generally agreed to within an order of magnitude. Differences were attributed to the conditional nature of the adsorption constant and the wide range of experimental conditions employed. However, the potentially important role of hydrous oxides of iron and manganese was highlighted. Photoassisted dissolution of a colloidal manganese oxide in the presence of fulvic acid has been studied by Waite *et al.* (1988). The movement and fate of hazardous materials in soil systems have been reviewed further by Kaufman and Hornick (1984).

The binding of PAHs to dissolved humic material has been

examined, using equilibrium dialysis and fluorescence techniques (McCarthy & Jimenez, 1985; Gauthier *et al.*, 1986). There was a direct relationship between the hydrophobicity of the PAH and the affinity for binding to dissolved humic material. The presence of humic substances has also been found to affect the formation of chlorinated PAHs (Johnsen & Gribbestad, 1988). Chlorinated PAHs were formed in lake water samples containing fluoranthene and benzo[*a*]pyrene, but none were detected in the presence of added humic substances.

5.4 MATHEMATICAL MODELS OF SOIL SYSTEMS

Dealing mainly with point source pollution, soil and groundwater models may be categorized into four conceptual groups. These are those that deal with the unsaturated soil zone (soil) and the saturated soil zone (groundwater), or they are geochemical or ranking models (Bonazountas, 1983). The first two are comparable in terms of mathematics and approach, the third deals with speciation chemistry, and the fourth relies on a screening system. Most soil models are concerned with vertical chemical flow and most groundwater models deal with horizontal flows. Three-dimensional models do exist (Narasimhan, 1975), but they are mathematically extremely complex and unwieldy in practice.

In general, models are classified as deterministic or stochastic. Deterministic models describe the compartment in terms of cause and effect relationships. They may be further classified into simulation models, which describe a system using verified empirical equations, and analytic models, in which derived equations describe the physico-chemistry of the system. Stochastic models incorporate measures of uncertainty such as risk and probability. Bonazountas (1983) suggested that deterministic or stochastic soil quality models combine two major modules:

1. The flow module or moisture module (or hydrologic cycle module), which predicts moisture behaviour.
2. The solute module, which predicts pollutant transport, transformation and soil quality.

Further information on the subject of soil and groundwater model parameters has been provided by Freeze and Cherry (1979), Bachmat *et al.* (1980), Bonazountas and Wagner (1981) and Mercer and Faust

(1981). The *Ground Water Modeling Newsletter* of the International Ground Water Modeling Center, Holcomb Research Institute, Butler University, Indianapolis, Indiana, is a useful source of information on recent developments in groundwater modelling.

5.4.1 Unsaturated soil zone (soil) modelling

Bonazountas (1983) categorized soil modelling into three different mathematical formulation patterns as follows:

A. Traditional differential equation (TDE) modelling.
B. Compartmental modelling.
C. Stochastic modelling.

The TDE moisture module of the model is derived from equations for water mass balance and water momentum, the Darcy equation and equations such as that for the surface tension of potential energy. Presented as a one-dimensional, vertical unsteady-state isotropic formulation, the resulting differential equation is

$$\partial[K(\psi)(\partial\psi/\partial_z+1)]/\partial_z = C(\psi)\,\partial/\partial t + S$$

and

$$V_z = -K(Z,\,\psi)\,\partial\theta/\partial_z$$

where Z is the elevation (cm), ψ is the pressure head (soil moisture tension head in the unsaturated zone) (cm), $K(\psi)$ is the hydraulic conductivity (cm min^{-1}), t equals time (min), S is the water source or sink term (min^{-1}), V_z is the vertical moisture flow (cm s^{-1}) and

$$C(\psi) = \mathrm{d}\theta/\mathrm{d} = \frac{\text{slope of the moisture } (\theta)}{\text{pressure head (cm}^{-1})}$$

and $\theta = Z + \psi$

The moisture module output provides the parameters V and θ as input to the solute module.

The TDE solute module is formulated with one equation describing pollutant mass balance in a representative soil volume $\mathrm{d}V = \mathrm{d}x\,\mathrm{d}y\,\mathrm{d}z$, as

$$\partial(\theta c)/\partial t = [\partial(\theta\,.\,K_\mathrm{D}\,\partial c/\partial z/\partial t] - [\partial(vc)/\partial t] - [\rho\,.\,\partial s/\partial t] \pm \sum P$$

where θ is the soil moisture content, C is the dissolved pollutant concentration in soil moisture, K_D is the apparent diffusion coefficient of compound in soil–air, v represents the Darcy velocity of soil moisture, ρ is the soil density, s is the adsorbed concentration of compound on soil particles, $\sum P$ is the sum of pollutant sources or sinks within the soil volume and z is the depth.

Further details of TDE-type soil models have been given by Mackay (1979), Enfield *et al.* (1980) and Schwartz and Growe (1980).

Compartmental soil models can be applied to both the solute and moisture modules. In the solute fate module, the law of chemical mass concentration is applied to a specific soil element over a specified interval. Either the entire soil matrix or its constituent phases are considered. Assuming equilibrium, the concentration in the other phase can be calculated. One of the most developed soil compartment models is the seasonal soil compartment model (SESOIL). This model is made up from a dynamic compartment moisture module and dynamic compartmental solute transport module (Bonazountas & Wagner, 1981). A schematic diagram of the soil matrix phases in SESOIL is presented in Fig. 5.4.

Stochastic models generally attempt to determine times taken to achieve overall concentration thresholds, rather than describe processes or concentrations in individual soil compartments. Stochastic models have been described by Schwartz and Growe (1980), Jury (1982), Schultz (1982) and Tang *et al.* (1982).

5.4.2 Saturated soil zone (groundwater) modelling

Groundwater modelling is based almost exclusively on TDE systems comprising flow and solute modules (Bachmat *et al.*, 1980). Analytical models result in mathematical expressions applicable to the entire groundwater compartment. In such models the only input requirements are averaged data for the entire compartment. Alternatively, numerical models formulate expressions applicable to the phases of the compartment, thus data are required for all phases in the model. Various numerical solution techniques can be found by reference to Bear (1979), Mercer and Faust (1981) and Prickett *et al.* (1981). An important consideration in groundwater models is that calibration is required for site-specific applications.

5.4.3 Ranking models

These are based on question–answer techniques and on specified weighting factors. Despite being subjective and of dubious scientific merit, their simplicity has led to their widespread use. Examples of such ranking models are those of JRB (1980), LeGrand (1980), ADL (1981), MITRE (1981) and that by Silka and Swearingen (1978).

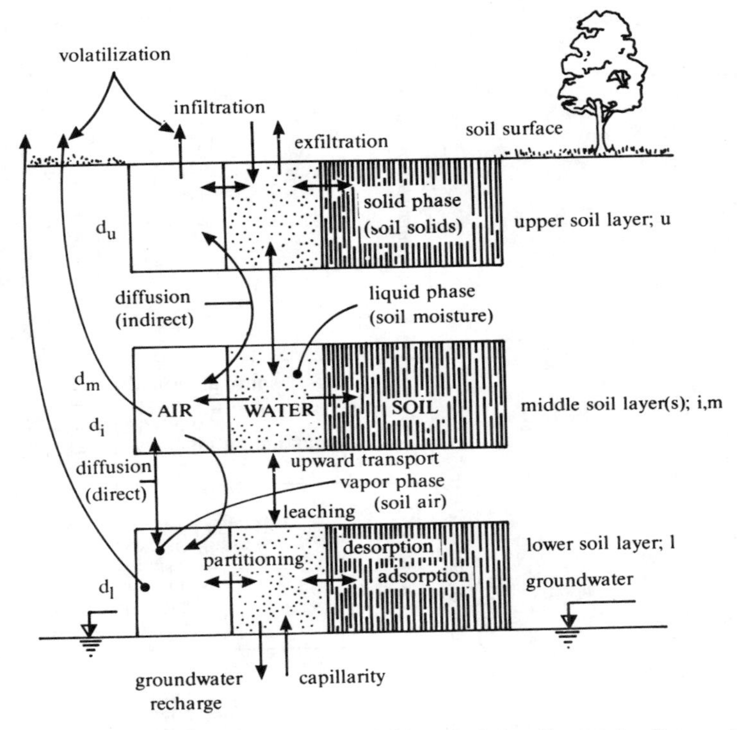

Fig. 5.4. Schematic of phases in SESOIL model soil matrix. Reproduced from Bonazountas and Wagner (1981).

5.4.4 Aquatic equilibrium models

Such models have been developed to detail specific processes within environmental phases. The models provide information on the distribution of various chemical species in specified situations. In the more popular equilibrium constant approach, a solution is obtained using an iterative procedure utilizing measured equilibrium constants for all mass action expressions (Nardstorm *et al.*, cited in Bonazountas & Fiksel, 1982). Details of GEOCHEM, an aquatic geochemical equilibrium model, were given by Sposito (1981*b*).

The alternative Gibbs free energy modelling approach is hampered by the limited number of free energy values and is therefore most useful for simple modelling systems. Tables 5.7 and 5.8 list and outline the main features of a variety of soil and groundwater models developed before 1984, while Table 5.9 summarizes the main features

Table 5.7
Selected models and features[a]

Model Acronym / Model Features	Model type[b]				Model formulation[c]					Mathematics[d]			Chemistry issues[e]					User concerns[f]				Contact/information
	Unsaturated Zone	Groundwater	Aquatic Equilibrium	Ranking	Flow Module	Solute Model	TDE Approach	Compartmental	Statistical, Other	Analytical	Numerical	Statistical	Organics	Inorganics	Metals	Gaseous Phase	Increased Chemistry	Input Data Requirements	Calibration	Level of Effort	Application Study	
PESTAN	*					*	*			*			*					L	L	L	*	Enfield *et al.* (1980)
SCRAM	*				*	*	*				*		*					H	H	H	*	Adams and Kurisu (1976)
SESOIL	*				*	*		*	*	*	*		*	*	*	*	*	M	M	M	*	Bonazountas and Wagner (1981)
AT123D	*	*				*	*			*			*	*	*			M	L	M		Yeh and Ward (1981)
PLUME		*				*	*			*			*		*			L	L	L	*	cited in Bonazountas (1983) as Wagner; OSU (405) 624–5280

PATHS		*				*	*		*	*	*	*	*	*			L	M	L	*	Nelson and Schur (1980)
MMT/VVT	*	*			*	*	*				*	*					H	H	H	*	Foote (1982)
FEM WASTE	*	*				*					*	*					H	H	H	*	Yeh and Ward (1981)
R. WALK		*				*	*				*	*					H	H	H	*	Prickett *et al.* (1981)
USGS Models	*	*			*	*	*				*	*	*	*	*		H	H	H	*	Appel and Bredehoeft (1978)
GEO CHEM			*			*	*	*			*		*	*		*	H	L	L	*	Sposito (1981*b*)
MITRE/JRB	*	*		*													M		M		MITRE (1981)/JRB (1980)
ADL/LeGRAND	*	*		*													M		M	*	ADL (1981)/LeGrand (1980)

[a] This is a partial list of available well-documented models. An EPA *Modeling Handbook and Catalogue* (Bonazountas & Fiksel, 1982) lists additional models.

[b] The use of complex (e.g. a numerical soil and groundwater package) models that can handle more than one compartment is not always desirable, since generalized packages tend to be cumbersome, unless especially designed (class of models).

[c] The most representative characteristics are given. The traditional differential equation (TDE) approach applies to the flow and solute module. Under 'other' are, e.g. linear analytical system solutions.

[d] The most representative characteristics are given since, e.g. statistical formulations can be subject to statistical, analytical or numerical solution procedures.

[e] Almost all models can simulate organic, inorganic and metal fate, assuming that a careful calibration via an adsorption coefficient may alter the model output to predict measured/monitored values. However, not all models have by design increased chemistry capabilities (e.g. cation exchange capacity; complexation), therefore, the most representative capabilities are indicated.

[f] L = low; M = medium; H = high input data requirements. In general, numerical models have higher input data requirements and calibration needs, therefore, may better represent spatial resolution of a domain. Compartment models provide an optimal compromise. The level of effort is intuitively defined here.

Adapted from Bonazountas (1983).

Table 5.8
Most important characteristics of major model categories

Model category	*Advantages*	*Disadvantages*	*Comments*
Soil and groundwater models			
TDE—Type	Clear formulation, known capabilities	Rigid model structure, limited capabilities	This has been the traditional computerized modelling approach
—Analytic	Easy model use; limited calibration possibilities; limited input data requirements; desk computer use	Rough averaged predictions of pollutant fate, limited application capabilities	To be used as overall fate (screening) tools
—Numerical	Wide range of applications; detailed spatial and temporal resolutions; increased chemistry capabilities	Extensive calibration requirements; input data intensive (nodes, elements, time); requires computer use and related skills	Recommended for site specific applications
Compartmental type	Can be tailored to user's requirements; increased chemistry capabilities; can better meet spatial and temporal domain requirements	Expected user interaction and problem understanding	Today's scientific tendency
Aquatic equilibrium model	Increased chemistry capability	Data intensive, parameters may not be available	Models at a developmental stage
Ranking models	Easy to use with available data	Simplistic approach, output reflects user's intuition	Employed by the EPA, US Army, Air Force and Navy

Data from Bonazountas (1983).

Table 5.9
Examples of post-1984 soil and groundwater modelling computer programs

Model	*Description*	*Reference*
MAGNUM-2D	Two-dimensional numerical model for transient or steady-state analysis of coupled heat transfer and groundwater flow in a fractured porous medium	England *et al.* (1985)
FEMTRAN-A	Finite-element program for numerical simulation of the two-dimensional transport of radionuclide decay chains through saturated and/or unsaturated sorbing porous media	Martinez (1985)
RAQSIM	Generalized flow model of a groundwater system using finite-element methods plus three support programs to compute recharge and discharge input data	Cady and Pekenpaugh (1985)
MODFLOW	New and final version of the USGS modular three-dimensional finite-difference groundwater flow model. Output can be fed to MODGRAF, a program developed by TECSOFT Inc. to provide graphics capability	McDonald and Harbaugh (1988)
SUTRA	(Saturated–unsaturated transport) A computer code which simulates fluid movement and the transport of either energy or dissolved substances in a subsurface environment	Voss (1988)
MOC	Developed by USGS to simulate solute transport in flowing groundwater. Graphics capability is provided by MOCGRAF, developed by TECSOFT Inc.	Konikow and Bredehoeft (1988)
INTERSAT	Solves one-, two- or three-dimensional models of groundwater flow. Packages are also available for surface water, water balance and evapotranspiration	Voorhees (1988)
INTERTRANS	A three-dimensional solute transport model, using the particle tracking technique to simulate contaminant advection and dispersion	Voorhees and Rice (1988)

of more recent models. An introduction to groundwater modelling with sample programs in BASIC has been published by Kinzelbach (1986).

APPENDIX

EQUATIONS IN MODELS USED TO COMPUTE VOLATILIZATION OF CHEMICALS FROM SOIL

1. Hartley (1969)—heat balance

$$f = \frac{\rho_{max}(1-h)}{\delta} \Big/ \left[\frac{1}{D_v} + \frac{\lambda^2 v \rho_{max} M}{kRT^2}\right] \tag{A1}$$

where f = flux of compound ($M\,L^{-2}T$); ρ_{max} = saturated vapour concentration at the temperature of the outer air ($M\,L^{-3}$); h = humidity of the outer air; δ = thickness of stagnant layer through which the chemical must pass (L); D_v = diffusion coefficient of vapour in the air ($L^2\,T^{-1}$); λ_v = latent heat of vaporization ($cal\,M^{-1}$); M = molecular weight ($M\,mol^{-1}$); k = thermal conductivity of air ($cal\,L\,K^{-1}$); R = gas constant ($cal\,mol^{-1}$); T = temperature (K) and where M, L, T in parentheses represents units of mass, length and time, respectively.

The second term in the denominator is a thermal component of the resistance to volatilization. It is significant primarily for water or quite volatile compounds; for less volatile compounds, this term can be ignored, resulting in the following simplified form:

$$f = \frac{D_v \rho_{max}(1-h)}{\delta} \tag{A2}$$

2. Hamaker (1972)

$$Q_t = 2C_0\sqrt{Dt/\pi} \tag{A3}$$

where Q_t = total loss of chemical per unit area over some time t ($M\,L^{-2}$); C_0 = initial concentration of chemical in soil ($M\,L^{-3}$); D = diffusion coefficient of vapour through the soil ($L^2\,T^{-1}$).

This equation may give low values since it ignores mass transfer due to the Wick effect. With the Wick effect, the chemical is transported from the soil body to the surface by capillary action. Its rate of evaporation is enhanced by the evaporation of the water causing the capillary action (Hartley, 1969).

Hamaker presented a second method based on total water loss—i.e. the loss due to vapour diffusion as well as the mass transfer of soil solution. The approximation for the loss of the dissolved, volatile chemical is

$$Q_t = \frac{P_{vp}}{P_{H_2O}} \frac{D_v}{D_{H_2O}} (f_w)_v + c(f_w)_L \tag{A4}$$

where f_w = loss of water per unit area ($M\,L^{-2}$); P_{vp} = vapour pressure of chemical; P_{H_2O} = vapour pressure of water; D_v = diffusion coefficient of chemical in air ($L^2\,T^{-1}$); D_{H_2O} = diffusion coefficient of water vapour in air ($L^2\,T^{-1}$); subscript V = loss of vapour; subscript L = loss of liquid; c = concentration of chemical in soil solution ($M\,M^{-1}$).

The use of this equation requires knowledge of the water flow in the soil as well as diffusion coefficients for water vapour and the chemical in question.

3. Mayer et al. *(1974)*

Diffusion is assumed to be the only mechanism supplying chemicals to the soil surface; therefore, the approach probably underestimates the volatilization rate. An analogy is drawn between the heat transfer equation (Fourier's law) and the transfer of matter under a concentration gradient (Fick's law). The equation is solved for various boundary conditions.

The one-dimensional diffusion equation, with a constant diffusion coefficient, D, is

$$\frac{\partial^2 c}{\partial z^2} - \frac{1}{D}\frac{\partial c}{\partial t} = 0 \tag{A5}$$

where c = concentration in the soil ($M\,L^{-3}$); z = distance measured normal to the soil surface (down is positive, surface is zero) (L); D = diffusion coefficient in soil ($L^2\,T^{-1}$); t = time (T).

Mayer *et al.* (1974) presented five different solutions, each applicable to a different set of boundary conditions.

4. Jury et al. *(1980)*

$$c_t = \rho_b c_s + \theta c_l + \eta c_g \tag{A6}$$

where c_t = total concentration in soil ($M L^{-3}$ soil); ρ_b = soil bulk density ($M L^{-3}$); c_s = adsorbed concentration ($M M^{-1}$); θ = volumetric soil water content ($L^3 L^{-3}$); η = soil air content ($L^3 L^{-3}$); c_l = concentration in the liquid phase ($M L^{-3}$ water); c_g = concentration in the gas phase ($M L^{-3}$ air).

The flux is

$$f = -D_g \frac{\partial c_g}{\partial z} - D_l \frac{\partial c_l}{\partial z} - f_w c_l \tag{A7}$$

where f = chemical flux ($M L^{-2} T^{-1}$); D_g = gas diffusion coefficient ($L^2 T^{-1}$); D_l = liquid diffusion coefficient ($L^2 T^{-1}$); f_w = water flux ($L^3 L^{-2} T^{-1}$).

The transport terms in the gas phase, $D_g(\partial c_g/\partial z)$, in the liquid phase, $D_l(\partial c_l/\partial z)$, and along with the soil water, $f_w c_l$, can be seen. The continuity equation is:

$$\frac{\partial c_t}{\partial t} + \frac{\partial f}{\partial z} = 0 \tag{A8}$$

These three equations can be solved to give concentrations in the soil column, along with the flux and total amount lost. The equations depend on a number of assumptions being fulfilled.

5. Dow Method (Swann et al., *1979)—surface application only*

For this method, relationships were determined between vapour pressure, water solubility and soil adsorption coefficient.

The volatilization rate was found to be proportional to the factor P_{vp}/SK_{oc} where P_{vp} is vapour pressure in mm Hg, S is water solubility in mg litre^{-1}, and K_{oc} is the soil adsorption coefficient in $(\mu g\, g^{-1})/(\mu g\, ml^{-1})$. Half-life for depletion of the chemical from the surface was found to be:

$$t_{1/2} = 1{\cdot}58 \times 10^{-8}\left(\frac{K_{oc}S}{P_{vp}}\right) \text{ days} \tag{A9}$$

and the rate constant for volatilization is

$$k_v = \frac{0{\cdot}693}{t_{1/2}} \text{ day}^{-1} = 4{\cdot}4 \times 10^7\left(\frac{P_{vp}}{K_{oc}S}\right) \text{ day}^{-1} \tag{A10}$$

The concentration of chemical on the surface at any time can be estimated from

$$C = C_o e^{-k_v t} \qquad \text{(A11)}$$

where C_o = concentration at $t = 0$ ($M\,L^{-3}$); k_v = volatilization rate constant (T^{-1}); t = time (T).

Chapter 6

The Aquatic Environment

6.1 INTRODUCTION

As soon as a chemical is released into the aquatic environment, it is subjected to an array of transport, transfer and transformation processes. Hydrodynamic transport is described in terms of advective and turbulent dispersive forces and occurs simultaneously with transfers to sorbed forms and irreversible transformation processes. In time, as a result of dilution, speciation and transformation, environmental concentration gradients of the chemical are established.

In this chapter, concerning the aquatic compartment of the environment, the physical processes of hydrodynamic transport, chemical solubility in water and volatilization from water are discussed. Molecular diffusion, of particular importance in modelling chemical behaviour in soil, is of much less consequence in the aquatic environment and is only mentioned in terms of its effect on volatilization processes. The chemical processes of ionization, hydrolysis, oxidation–reduction and complexation have all been considered in the previous chapter and are not reiterated. However, problems of variable metal speciation are raised, and the process of photolysis is highlighted. The water requirement of living organisms is acknowledged by outlining in this chapter the importance of biodegradative processes on chemical fate and persistence.

Mathematically, the parallel processes of transport, transfer and transformation can be described in terms of their total effect on the rate of change of chemical concentration. This environmental rates approach may be achieved by a set of coupled differential equations and is discussed in some detail. Finally, the major alternative to such a modelled process, the model ecosystem approach, is considered with reference to existing models.

6.2 THE HYDROSPHERE

Of the $5{\cdot}10 \times 10^8\,km^2$ of the Earth's surface area, 71%, or $3{\cdot}61 \times 10^8\,km^2$, consists of water. The major proportion is held in the oceans and icecaps, followed by groundwater, lakes and rivers. Only a minute fraction is retained in the atmosphere. Table 6.1 presents the volumes and average residence times of the five major aquatic environmental reservoirs.

Ocean currents have created a partial barrier at the equator, which effectively divides the oceans into two major compartments. The water area north of the equator is estimated at $154 \times 10^6\,km^2$, and south of the equator, $209 \times 10^6\,km^2$ (Neely, 1980). Ocean mixing zones have been determined in studies such as the Geochemical Ocean Section Programme (Hammond, 1977), using data from ocean tritium concentrations. More recently, the effects of shear and surface boundary on contaminant diffusion in the ocean were discussed by Gabric (1983). Figure 6.1 summarizes current concepts of major ocean processes for reactive elements.

Table 6.1
Volume of water in the major environmental compartments and average residence time for each compartment

Compartment	*Volume* ($\times 10^3\,km^3$)[a]	*Percentage of total*	*Residence time (years)*[b]
Oceans	1 320 000	97	3 000
Ice caps and glaciers	28 000	2	8 000
Ground water	9 120	0·6	330
Lakes and rivers	228	0·016	7[c]
Atmosphere	13	0·001	0·027

[a] Neely (1980), Chapter 5.
[b] Lvovitch (1977).
[c] Rivers have a residence time of 0·03 year.

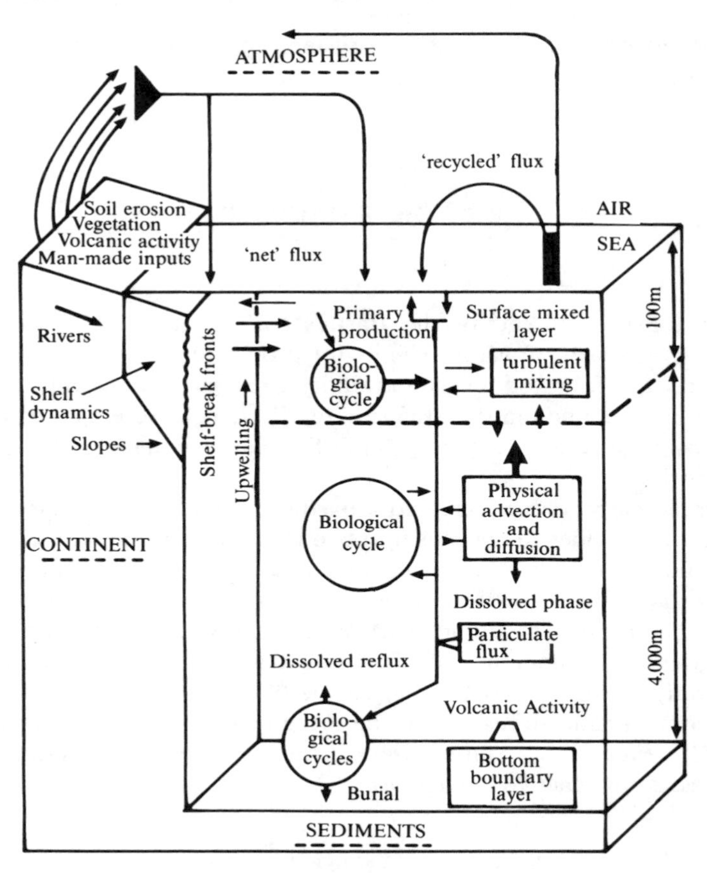

Fig. 6.1. Schematic diagram of the major ocean processes for reactive elements. Reproduced from Buat-Menard (1983).

6.3 PHYSICAL PROCESSES

6.3.1 Hydrodynamic transport

Hydrodynamic transport is usually modelled in terms of the combined forces of advection and dispersion. Advection may be considered to be the flow of water through a system, and dispersion is a statistical accounting of the effects of turbulence, storm surges, internal waves and other phenomena not amenable to detailed mechanistic description (Burns, 1983). Hydrodynamic studies have been outlined for rivers by Shen (1979), for estuaries by Nihoul (1978) and Baumgartner and Callaway (1972), and for lakes and oceans by Gibbs (1977).

Physical transport processes and terms were coupled with a gross biological compartment by Bostater *et al.* (1981) in order to provide a more meaningful fate model for di-*n*-octylphthalate (DNOP) in a subestuary of Chesapeake Bay. It was argued that organic chemicals can be transported in both the dissolved phase and the particulate phase, adsorbed to surface silts and clays. Thus, any general model of toxic chemical transport in estuaries had to calculate the transport of dissolved substances, the transport of small particulate substances and the partitioning of the chemical between the two phases.

Advection

Advection was divided into a net, freshwater flow component and a tidal flow component. In geometrically simple, well-mixed estuaries, the instantaneous flow at a cross section is the sum of these two components. River bends and lateral depth variations add a large-scale lateral circulation component to the flow, generally augmenting downstream transport in the channel and upstream transport in the shallows. Vertical and longitudinal density gradients can set up two-layer flows, causing net upstream transport of relatively saline bottom waters and net downstream transport of relatively fresh surface waters. Other forces, such as wind stress and Coriolis acceleration, often contribute significantly to advective flows. The product of the resultant flow and the concentration at a location gives the advective transport.

Particle transport

Because many organic chemicals bind with aquatic particulate matter, particle transport may determine the fate of a compound. One strategy for environmental modelling is to 'piggy-back' the transport of sorbed chemicals on a model of the transport of particulates. Clay-sized particles with very low settling velocities are transported by fresh water in suspension along with dissolved substances. When low-salinity concentrations are first encountered, however, the mutually repulsive surface charges of the clays are neutralized. Clay aggregates with higher settling velocities then form in a region known as the turbidity maximum. These settling flows tend to be transported upstream in the bottom waters, which leads to the formation of sediment deposits at the landward boundary of the salinity intrusion. Above this point, fresh water causes disaggregation and resuspension of unconsolidated clay deposits. The result is a fine 'sediment trap' at the turbidity maximum. The location of this region in an estuary is controlled by the salinity concentrations and the properties of the predominant clay. The actual rates of flocculation, setting and erosion of these clays depend on a complicated set of processes that are affected by environmental conditions. They are subjected to large variability over the tidal cycle and the spring–neap cycle. Settling speeds of sewage sludge in seawater have been reviewed by Lavelle *et al.* (1988).

The partitioning of copper and zinc to suspended solids in the Flint River, Michigan, has been discussed by McIlroy *et al.* (1986). The reversible component partition coefficient for nickel and cobalt sorbed to montmorillonite and quartz has been shown to be a function of particle concentration (Di Toro *et al.*, 1986), and copper and cadmium uptake by estuarine sedimentary phases has been modelled by Davies-Colley *et al.* (1984). Dissolved organic compounds in an estuarine environment have been found to have a relatively high binding capacity for copper, with binding capacity directly proportional to dissolved organic carbon concentrations (Newell & Sanders, 1986). The whole area of metal scavenging by sorbent particles has been comprehensively reviewed by Honeyman and Santschi (1988).

Estuarine model

To help assess the various transport processes, equations were developed to express transport in terms of measurable variables linked by basic laws and empirical relationships (Bostater *et al.*, 1981). Physical, chemical and biological transformations were accounted for with the inclusion of a source or sink term S (Lassiter *et al.*, 1979); the general form of this term is the same as that currently being used in ecological compartment models of the form:

$$S = -Vo - Pd - Ps - Hy - Au - Md - Fu + Fe \pm Se + L$$

where Vo equals volatilization, Pd equals direct photolysis, Ps equals sensitized photolysis, Hy equals hydrolysis, Au equals breakdown by photoautotrophs, Md equals microbial degradation, Se equals exchange with sediments, and L represents direct loadings of toxicant.

Others have added to S, terms representing uptake and excretion by fish (Fu, Fe) (Neely & Blau, 1977). All of this work has been done in the context of compartmental analysis, where the transport equation is reduced to that for a single, continuously stirred tank reactor where

$$\frac{V\,\mathrm{d}c}{\mathrm{d}t} = VS$$

where V is the volume of the compartment (m^3).

Inflows and exports are included in S as specified mass fluxes. Only expense prevents the direct use of the compartmental model S terms in the more complicated transport equations. Estuarine models differ most from pond models in the complexity represented in the sediment exchange term, Se. The extent to which these terms can be aggregated into an overall 'summation rate content' varies with the complexity of the estuary and the toxic substance.

Table 6.2 lists the chemical properties of di-*n*-octyl phthalate ester (DNOP) and 'average' characteristics of the Chester River

Table 6.2
Model DNOP parameters

	Constant description	*Units*	*Value*
KOW	octanol–water partition coefficient	g water g^{-1} fat	$1{\cdot}6 \times 10^4$
OCC	lipid fraction of oysters	g fat g^{-1} edible	0·02
KO	effective oyster–water partition coefficient	g water g^{-1} edible	320
BM	benthic oyster biomass	g edible m^{-2}	1·0
KPC	organic sediment–water partition coefficient	g water/organism–sediment	$5{\cdot}9 \times 10^4$
OC	organic fraction of sediments (wet basis)	g organism g^{-1} sediment	0·01
KP	effective sediment–water partition coefficient	g water g^{-1} sediment	590
SS	suspended sediment	g m^{-3}	30
KOH	alkaline hydrolysis rate	litre/molar/day	27
pH	hydrogen ion activity	standard units	7·7
KH	effective hydrolysis rate	litre day^{-1}	1×10^{-5}
KM	microbial degradation rate	litre/organism day^{-1}	$7{\cdot}4 \times 10^{-12}$
BP	sediment bacterial population	organism ml^{-1}	$1{\cdot}0 \times 10^6$
BPOP	effective depth-average bacterial population	organism ml	10^4 to 10^5
KB	effective bacterial degradation rate	litre day^{-1}	$2–5 \times 10^{-4}$
KD	effective decay rate	litre day^{-1}	$2–5 \times 10^{-4}$

Data from Bostater *et al.* (1981).

environment estimated for the kinetic subroutines. It was assumed that bottom sediments exchange freely with suspended sediments; that the particulate-adsorbed and biologically absorbed phases of DNOP are in continuous equilibrium with dissolved DNOP; and that DNOP is lost only through advective and dispersive transport, alkaline hydrolysis and microbial decay. Transport equations were solved for total DNOP with the following sink term:

$$S = -(KB + KH)Cw$$

where Cw equals dissolved DNOP (mg m^{-3}). Also

$$KB = KM \times BPOP$$

and

$$KH = KOH \times 10^{(\mathrm{pH}=14)}$$

The total DNOP (Ct) may be partitioned into dissolved (Cw), particulate (Cp) and biological (Cb) phases (all in mg m^{-3}) with the following equations:

$$Cw = \frac{Ct}{1 + KP \times SS \times 10^{-6} + KO \times BM \times 10^{-6}}$$

$$Cp = \frac{Ct \times KP \times SS \times 10^{-6}}{1 + KP \times SS \times 10^{-6} + KO \times BM \times 10^{-6}}$$

$$Cb = \frac{Ct \times KO \times PM \times 10^{-6}}{1 + KP \times SS \times 10^{-6} + KO \times BM \times 10^{-6}}$$

The concentrations per unit of sediment, Gp, and per unit of biomass Gb (oysters), both in mg kg^{-1}, can be calculated from the volumetric concentrations as follows:

$$Gp = \frac{Cp}{(SS \times 10^{-6})}$$

$$Gb = \frac{Cb}{(BM \times 10^{-6})}$$

Parameter sensitivity analysis suggested that accurate prediction of the total DNOP throughout the Chester River demands a thorough understanding of its dispersive transport and boundary exchanges. Accurate prediction of DNOP concentrations in oysters calls for an understanding of the partitioning process, including both the octanol–water partition coefficient and the lipid content of the oysters.

6.3.2 Solubility in water

Water solubility is one of the most significant factors affecting the fate and transport of chemicals in the environment. Chemicals that are highly soluble are quickly distributed by the hydrologic cycle, tend to have relatively low soil adsorption coefficients, and bioconcentrate only slightly in aquatic biota. Such compounds are also readily biodegradable by microorganisms (Lyman, 1982*a*).

Chemical solubility in water is defined as the maximum amount of the substance that will dissolve in pure water at a specified temperature. With a solid or liquid organic chemical, if this concentration is exceeded, a biphasic situation will evolve, consisting of a saturated aqueous solution and a solid or liquid organic phase. Solubility may be affected by temperature, salinity and the presence of dissolved organic matter. The water solubility behaviour of seven medium molecular weight, liquid binary hydrocarbon mixtures was studied by Burris and MacIntyre (1985). It was found that the ratio of concentrations of hydrocarbon solutes was independent of the degree of saturation of the binary hydrocarbon mixture with water, the implication being that equilibrium phenomena observed could be applicable under nonequilibrium conditions. The capability of ideal absorbed solution theory to predict multicomponent competitive interactions between volatile organic chemicals was tested for chloroform, bromoform, trichloroethene, tetrachloroethene, 1,2-dibromoethane and chlorodibromomethane by Crittenden *et al.* (1985). Predictions were tested by measuring the degree of removal of the chemicals by adsorption on to granular activated carbon and were satisfactory for the 256 isotherm data collected. Error analysis demonstrated that under certain simplifying assumptions, the Freundlich equation gave an adequate representation of the data.

A variety of solubility estimation methods exist; e.g. the use of regression equations linking the octanol–water partition coefficient (Fig. 6.2) or melting point with solubility (Coates *et al.*, 1985; Miller *et al.*, 1985). Partition coefficients may be estimated from structure, and the solubility of hydrocarbons and halocarbons may be estimated by the addition of atomic fragments. Solubility at any temperature can be calculated from theoretical equations incorporating estimated activity coefficients (Opperhuizen *et al.*, 1988; Banerjee & Howard, 1988). Figure 6.3 illustrates some of the available routes to the estimation of aqueous solubility for organic chemicals. Comprehensive details of such estimation methods have been given by Lyman (1982*a*).

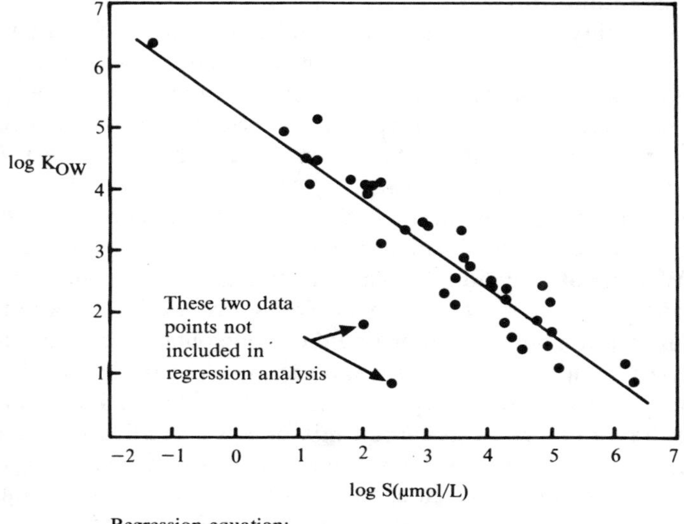

Regression equation:

$$\log S = -1\cdot 12 \log K_{OW} + 7\cdot 30 - 0\cdot 15\, t_m \ (\mu\text{mol/L})$$

$$r^2 = 0\cdot 992$$

t_m = 25°C if compound is liquid at that temperature

Fig. 6.2. Correlation between solubility (S) and octanol/water partition coefficient (K_{OW}). Forty-one mixed compounds including aromatics and chlorinated hydrocarbons. Reproduced from Banerjee *et al.* (1980).

6.3.3 Volatilization from water

The vaporization of organic chemicals from water bodies is a significant mass-transfer pathway from water to air. Volatilization is controlled by the chemical's solubility, molecular weight and vapour pressure, and by the nature of the air–water interface through which it must pass. Volatilization rates vary widely and are affected by a variety of known, but rapidly changing factors. Atmospheric conditions, particularly wind speed and stability, also affect the rate-controlling factors. These factors are often nonlinearly interdependent and do not behave in simple deterministic ways (Thomas, 1982*b*).

The theory that resistance to mass transport exists in both the gas and liquid phase interfacial layers was first discussed by Whitman (1923). However, it was not until 50 years later that most empirical work was carried out. From these studies, four basic approaches

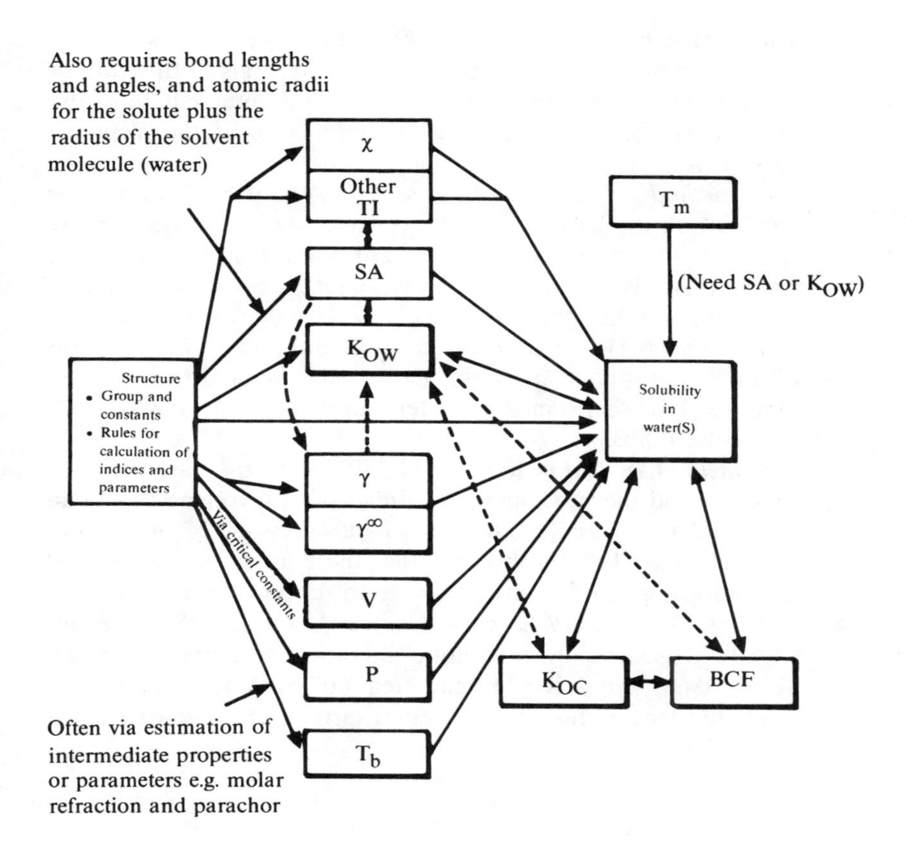

Fig. 6.3. Pathways for estimating the aqueous solubility of organic chemicals. Reproduced from Lyman (1982). BCF, Bioconcentration factor; K_{OC}, Soil-sediment adsorption coefficient; K_{OW}, Octanol-water partition coefficient; P, Parachor; SA, Molecular surface area; T_b, Boiling point; T_m, Melting point; TI Topological indices; V, Molar volume; γ, Activity coefficient in water; γ^{∞}, Infinite dilution activity coefficient in water; χ, Molecular connectivity parameter. $\rightarrow$, Pathway (unidirectional) leading to S; ⇢, Other pathways of interest; ⇄, Bidirectional pathways; □, Properties that may be measured in the laboratory (directly or indirectly).

emerged, namely:

1. *The method of Mackay and Wolkoff (1973)*—expresses the flux from a solution in water to the air above in terms of the ratio of contaminant mass in the vapour phase to the total vapour of the water plus chemical, expressed as a function of the chemical vapour pressure.
2. *The method of Liss and Slater (1974)*—volatilization was analysed on the basis of a two-layer film. The main water body is assumed to be well mixed with a thin layer on the surface in which there is a concentration gradient. The same situation exists for the air phase. At the interface between these two layers (Fig. 6.4) is a concentration discontinuity, and the ratio of air to water concentrations across it is assumed to equal the Henry's Law constant. Transfer through these films is by molecular diffusion.
3. *The method of Chiou and Freed (1977, 1979)*—based on gas dynamic and thermodynamic considerations involving the mean free path of molecules and the vapour pressure of the chemicals. Thomas (1982*b*) observed that the efficacy of this method is unknown as no data are available to support its validity.
4. *The method of Smith and Bomberger (1979) and Smith* et al. *(1980)*—based upon reaeration studies by Tsivoglou *et al.* (1965, 1968) and Tsivoglou and Neal (1976). Volatilization rate is related by a diffusivity correction term to the rate of oxygen reaeration.

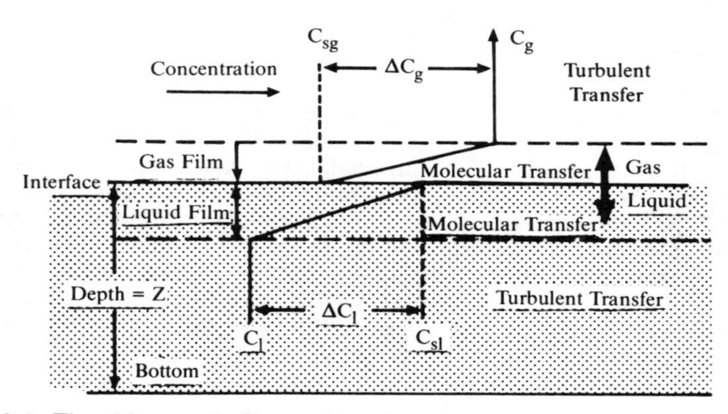

Fig. 6.4. The Liss and Slater (1974) two-layer model of gas–liquid interfaces. Reproduced from Thomas (1982).

The method for estimating volatilization from water recommended by Thomas (1982*b*) followed the two-film concept of Liss and Slater (1974). The method was based on a finite difference approximation to Fick's law of diffusion which can be written as

$$N = kC \tag{1}$$

where N is the flux (g cm^{-2} s^{-1}), k equals D/z, a first-order exchange constant (cm s^{-1}), D being the coefficient of molecular diffusion of chemical in the film (cm^2 s^{-1}) where z is the film thickness (cm); C is the concentration difference across the film (g cm^{-3}).

In a steady-state process, eqn (1) becomes

$$N = k_g(C_g - C_{sg}) = k_1(C_{s1} - C_1) \tag{2}$$

where K_g is the gas-phase exchange coefficient (cm s^{-1}), C_g is the concentration in the gas phase at the outer edge of the film (g cm^{-3}), C_{sg} is the concentration in the gas phase at the interface (g cm^{-3}), k_1 represents the liquid-phase exchange coefficient (cm s^{-1}), C_{s1} is the concentration in the liquid phase at the interface (g cm^{-3}), and C_1 is the concentration in the liquid phase at the outer edge of the film (g cm^{-3}).

The nondimensional Henry's Law constant (H') relates the concentration of a compound in the gas phase to its concentration in the liquid phase:

$$H' = C_{sg}/C_{s1} \tag{3}$$

Equation (2) can then be written as

$$N = \frac{C_g - H'C_1}{1/k_g + H'/k_1} = \frac{C_g/H' - C_1}{1/k_1 + 1/H'k_g} \tag{4}$$

The overall mass transfer coefficients for the gas phase (K_G) and liquid phase (K_L) can be defined as follows:

$$1/K_G = 1/k_g + H'/k_1 \tag{5}$$

and

$$1/K_L = 1/k_1 + 1/H'k_g \tag{6}$$

By substitution in eqn (4):

$$N = K_G(C_g - H'C_1) = K_L(C_g/H' - C_1) \tag{7}$$

The Henry's Law constant can also be written in the form:

$$H = P_{vp}/S \tag{8}$$

where P_{vp} is in atm, S is in mol m^{-3} and H is in atm m^3 mol^{-1}.

Figure 6.5, a graphical representation of eqn (8) shows values of H for water, air and numerous organic compounds. Figure 6.6 shows that H values for different chemicals give some insight into the controlling rate process and presents some generalizations concerning the volatility of chemicals that fall within postulated H ranges.

Mackay and Shiu (1975) pointed out that for materials that are solid at the temperature in question, the correct vapour pressure is the vapour pressure of the material in a hypothetical liquid state. This may be estimated by interpolating the vapour pressure/temperature curve to below the triple point junction of solid, liquid and vapour states. If the vapour pressure of the solid is used in the calculation, a lower value of H will be estimated than what is actually observed. Similarly, if the substance has a vapour pressure greater than 1 atm, then the vapour pressure should be estimated by extrapolation beyond the critical point.

Henry's Law constant for seven aldehydes have been determined as a function of temperature by bubble-column and head-space techniques (Betterton & Hoffman, 1988), while Fendinger and Glotfelty (1988) have described the use of a wetted-wall column to determine H for lindane (γBHC), alachlor and diazinon. The temperature dependence of Henry's Law constants has been established for trihalomethanes in water, which in the range 10–1000 μg litre^{-1} were independent of concentration (Nicholson *et al.*, 1984). Rates of volatilization from water have been reported for tetrachlorodibenzodioxin and ethylene dibromide by Podoll *et al.* (1986) and Rathbun and Tai (1986), respectively.

Henry's Law constants for polychlorinated hydrocarbons, particularly PCBs, have been the subject of numerous studies, using a variety of predictive and experimental techniques (Burkhard *et al.*, 1985*a,b*; Arbuckle, 1986, Burkhard *et al.*, 1986; Gossett, 1987; Dunnivant *et al.*, 1988). Table 6.3 presents a summary of published data for the 17 PCB congeners studied by Dunnivant *et al.* (1988). Differences between H values reported by various authors were attributed to differences in the method of data analysis and difficulties in measuring total initial aqueous PCB concentrations.

When crude oils are spilled on land or water, volatilization is often a

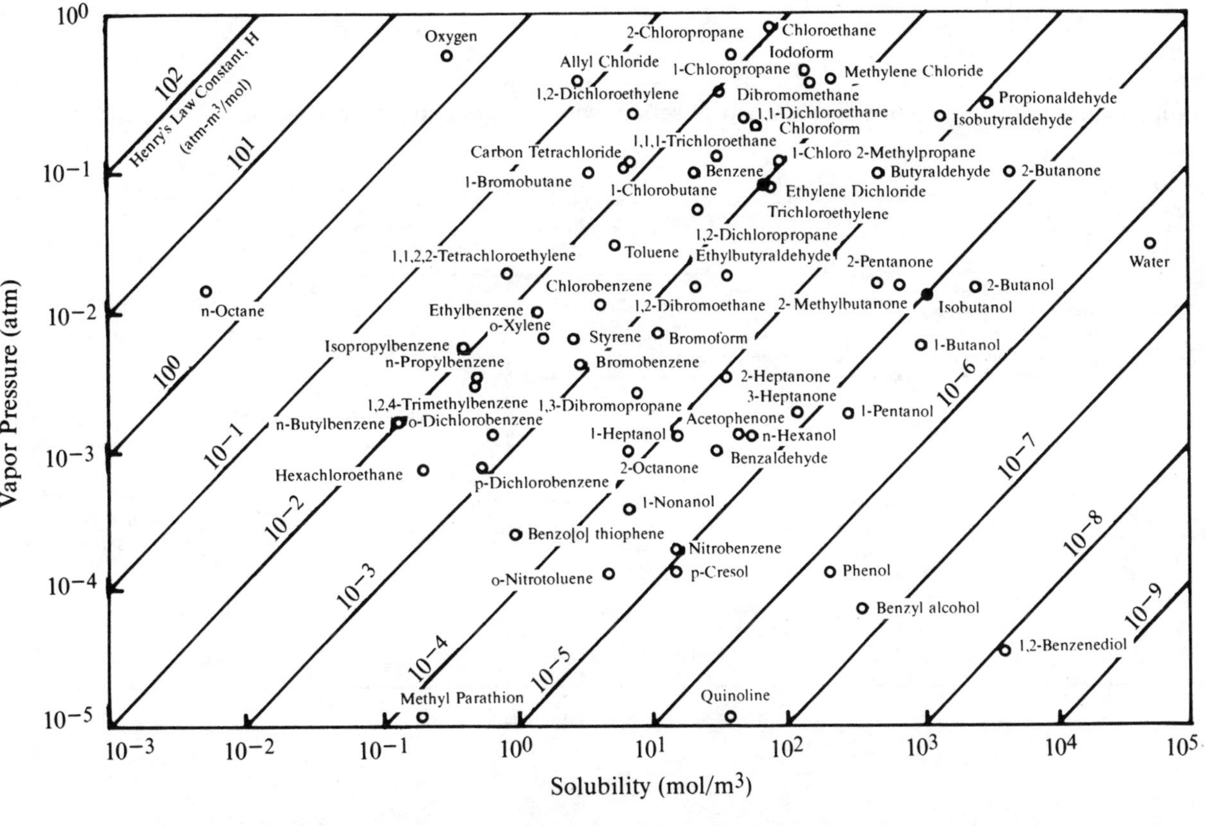

Fig. 6.5. Solubility, vapour pressure and Henry's Law constant for selected chemicals. Data from Mackay and Yuen (1979). Figure reproduced from Thomas (1982).

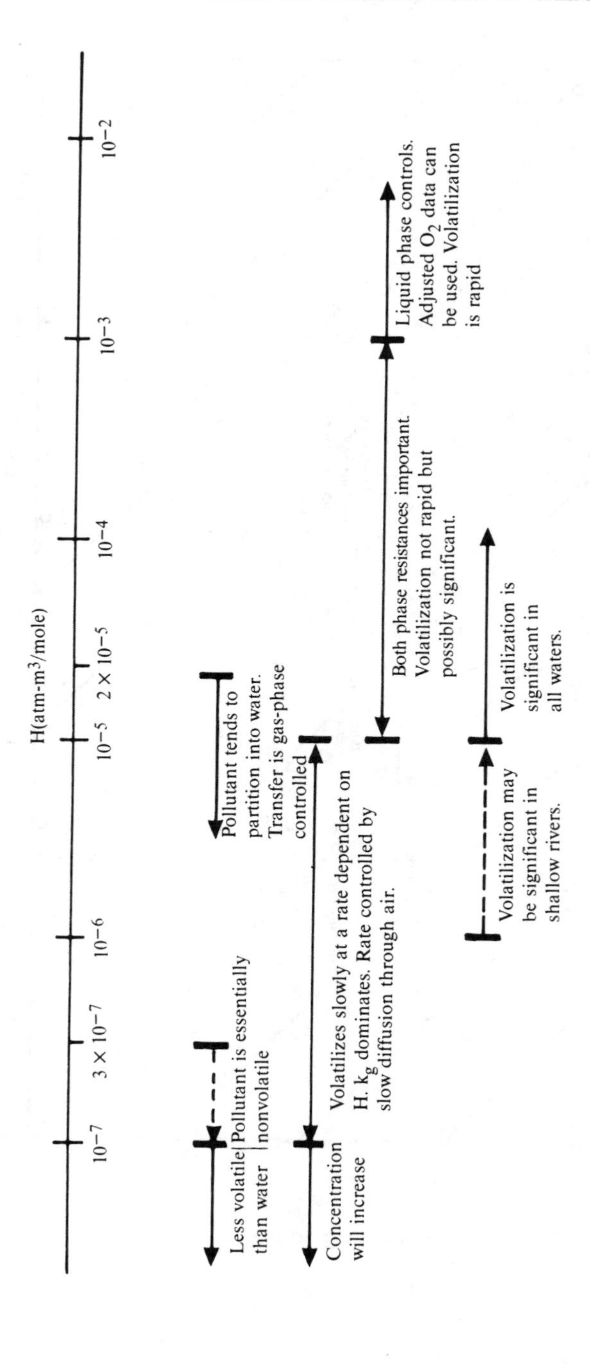

Fig. 6.6. Volatility characteristics associated with various ranges of Henry's Law constant (*H*). Reproduced from Thomas (1982).

Table 6.3
A summary of published data for Henry's Law constants (±SD) for 17 polychlorobiphenyl congeners

IUPAC no.	*Biphenyl chlorine substitution pattern*	*Henry's law constant* ($\times 10^4$ *atm* m^3 mol^{-1})						
		Dunnivant[a] *(25°C)*	*Coates*[b] *(25°C)*	*Oliver*[c] *(20°C)*	*Hassett*[d] *(25°C)*	*Murphy*[e] *(room temp.)*	*Westcott*[f] *(25°C)*	*Burkhard*[g] *(25°C)*
4	2, 2′	3·37 (0·267)				2·2		5·5
9	2, 5	3·88 (0·075)						3·27
11	3, 3	2·33 (0·106)						1·34
12	3, 4	2·05 (0·086)						0·948
15	4, 4′	1·99 (0·110)	1·45 (0·14)			≈3		1·09
26	2, 3′, 5	3·25 (0·095)						2·80
30	2, 4, 6	6·49 (0·246)						3·68
40	2, 2′, 3, 3′	2·02 (0·050)		1·2				2·00
52	2, 2′, 5, 5′	3·42 (0·086)		1·2	0·25	2·6	3·1–5·3	5·25
54	2, 2′, 6, 6′	5·50 (0·230)	1·48 (0·07)					18·60
53	2, 2′, 5, 6′	4·06 (0·125)						2·56
77	3, 3′, 4, 4′	0·94 (0·076)						0·431
101	2, 2′, 4, 5, 5′	2·51 (0·163)		0·7			1·1–3·5	3·23
104	2, 2′, 4, 6, 6′	8·97 (0·432)						18·30
128	2, 2′, 3, 3′, 4, 4′	0·302 (0·023)				5		0·676
153	2, 2′, 4, 4′, 5, 5′	1·32 (0·107)	1·23 (0·01)	0·6		3·5		1·77
155	2, 2′, 4, 4′, 6, 6′	7·55 (0·940)	1·15 (0·16)					15·50

[a] Dunnivant *et al.* (1988).
[b] Coates (1984).
[c] Oliver (1985).
[d] Hassett and Milicic (1985).
[e] Murphy *et al.* (1983).
[f] Westcott *et al.* (1981).
[g] Burkhard *et al.* (1985).
Reproduced from Dunnivant *et al.* (1988).

significant process of mass loss from the spill. The volatilization rate controls the atmospheric vapour concentration and so influences the toxic threat from inhalation in the vicinity of the spill. The evaporation of biologically toxic molecular weight fractions may modify the density, viscosity and toxicity of the residue. Consequently, the paper by Stiver and Mackay (1984) is of particular interest, since the latter authors developed the theoretical background to quantifying volatilization rates of crude oils, and related this to three experimental techniques. These were tray evaporation, gas stripping and distillation. The concept of a dimensionless evaporative exposure was devised, providing characterization of environmental or laboratory conditions. It incorporated the area of the oil, its volume, the prevailing wind speed and the time of exposure. Measured and calculated evaporation losses of two petroleum hydrocarbon herbicide mixtures have been reported by Woodrow *et al.* (1986).

Hill *et al.* (1976) investigated the behaviour of vinyl chloride in aquatic systems under conditions simulating large inputs of the monomer from industrial sources. An initial conceptual model considered some of the components and reactions that need to be evaluated in preliminary modelling exercises. Experiments showed that volatilization was the only process whose rate was not essentially zero. Thus, the model could be much simplified to an atmospheric and aquatic two-compartment system, with vinyl chloride escaping to the atmosphere by the single process of volatilization. The only other data required were the chemical's input and output rates to the aquatic compartment.

6.4 CHEMICAL PROCESSES

6.4.1 Aqueous photolysis

Photochemical processes may affect the fate of chemicals in aqueous environments both by *direct* photolysis, in which the substance itself absorbs solar radiation, and by indirect or *sensitized* photolysis, in which energy is transferred from some other species. Rates of photolysis depend on properties of the aquatic environment such as the intensity and spectrum of insolation and the presence or absence of sensitizers and quenchers. Equally important is the extent of light absorption and the inherent tendency of the chemical to undergo photochemical reaction (Harris, 1982*b*).

Factors common to all natural water systems, be they freshwater or marine, include the widespread involvement of unidentified organic chromophores (light-absorbing entities) in processes involving either net redox reactions, energy transfer or both. Dissolved oxygen is universally important as an acceptor of energy or electrons and as a participant in secondary reactions. Also common are secondary free radical reactions of organic and inorganic species and a much higher incidence of direct photolysis of pollutants than of natural molecules of low molecular weight and known structure (Zafiriou *et al.*, 1984).

However, there are many significant differences between freshwater and saltwater environments, not least because illuminated zones of freshwater are generally much shallower and more variable than those of oceanic central regions. Concentrations of dissolved organic compounds and suspended particulates, whilst being variable, are usually one to four orders of magnitude lower in seawater than in freshwater. The pH and ionic strength of seawater is $8 \pm 0{\cdot}5$ and 0·7 respectively, compared with 7 ± 3 and 0·001 to 3 for freshwater. Also, the anionic composition of seawater is dominated by chloride and sulphate, and again while freshwaters may vary, in the main they are dominated by bicarbonates.

The photochemistry of natural waters has been succinctly reviewed by Zafiriou *et al.* (1984) who have also compiled a comprehensive bibliography on the subject, which is available as a separate publication (Zafiriou, 1984). Photocatalysed dehalogenation of hydrocarbon contaminants in water has been reviewed by Ollis (1985), and environmental phototoxicity by Larson and Berenbaum (1988). The photochemistry of petroleum in water has been considered by Payne and Phillips (1985).

Direct photolysis

In this process, absorption of a photon promotes a molecule from ground to an electronically excited state. The excited molecule then either reacts to yield a photoproduct or decays to its ground state. The efficiency of these energy conversions is measured in terms of 'quantum yield'. The law of conservation of energy demands that primary quantum efficiencies sum to unity; therefore, photochemical reactivity is composed of two factors, the absorption spectrum and the quantum efficiency. At present, there are no methods for estimating the quantum yield of an organic chemical (Harris, 1982*b*).

Only electronic transitions, corresponding to ultraviolet–visible light absorption, are inherently energetic enough to lead to chemical reactions. The regime of importance for photochemical transformations is thus confined to light with a wavelength of 110–750 nm. In the environment, the stratospheric ozone layer effectively prevents radiation of less than 290 nm from reaching the ecosphere, further narrowing the effective wavelength band. The excitation process is expressed as

$$P \xrightarrow{h\nu} P^*$$

where P is the ground state molecule, $h\nu$ is a quantum of light and P^* represents an excited-state molecule.

In quantitative terms, the light absorbed in this process is given by the Beer–Lambert law (Dyer, 1965) as

$$A = \log \frac{I_0}{I} = \varepsilon c l$$

where A is the absorbance, I_0 is the intensity of incident light (of specified wavelength), I is the intensity of transmitted light (of same wavelength), ε is the molar absorptivity (extinction coefficient) of the absorbing species, c is the concentration of absorbing species and l is the depth of the absorbing medium.

The product of light intensity, chemical absorptivity and reaction quantum, when integrated across the solar spectrum, yields a pseudo first-order photochemical transformation rate constant (Burns, 1983). Aquatic environment photolysis half-lives, calculated as a function of season, latitude, time of day, water body depth and ozone layer thickness, have been discussed further by Zepp and Cline (1977).

Decreases in direct photolysis rates with increasing depth seem to be more pronounced for PAHs that absorb shorter wavelengths than for those that absorb light at relatively longer wavelengths (Zepp & Schlotzhauer, 1979). Furthermore, while photoreactions of many PAHs seem to be relatively rapid throughout the upper (35 m) mixed layer of oceans, photolysis is greatly slowed in turbid water because of light attenuation and partitioning of aromatics on to suspended particulate matter and bottom sediments. Photolysis rates in turbid water are particularly reduced for those higher molecular weight aromatics having the greatest tendency to absorb on to suspended particular matter (Payne & Phillips, 1985). Using 313-, 366- and 436-nm wavelength light, Zepp and Schlotzhauer (1979) calculated near-surface half-lives for direct photochemical transformations of 13

PAHs at 40°N latitude (midday, summer) to be as follows:

naphthalene (71 h)
2-methylnaphthalene (54 h)
1-methylnaphthalene (22 h)
fluororanthene (21 h)
phenanthrene (8·4 h)
chrysene (4·4 h)
anthracene (0·75 h)
pyrene (0·68 h)
benz[*a*]anthracene (0·59 h)
benzo[*a*]pyrene (0·54 h)
9,10-dimethylanthracene (0·35 h)
9-methylanthracene (0·13 h)
naphthacene (0·034 h)

Half-lives computed for photosensitized oxidation of most of the aromatics were many orders of magnitude longer, and several reactions proceeded in the absence of molecular oxygen. The authors concluded that singlet oxygen accounts for only a small fraction of the photoreaction of *dissolved* PAHs in water.

Mill *et al.* (1981), using methods of Zepp and Cline (1977), calculated rate constants and half-lives of the components of eight PAHs in water undergoing photolysis in sunlight as a function of season at 40°N latitude. The quantum yield, ϕ, was calculated from the relation:

$$\phi = k_p / 2{\cdot}3 \varepsilon I_0 r$$

where ε is the molar absorptivity, I_0 is the solar irradiance and r is a reactor constant.

Experimental rate constants were generated using the relation:

$$k_{pE} = 2{\cdot}3\phi \sum I_{0\lambda} \varepsilon_\lambda$$

where k_{pE} is the first-order rate constant (s^{-1}) and $I_{0\lambda}$ is the solar irradiance in surface water (einstein lx s^{-1}) at each wavelength at which absorption is measurable.

Limitations to this approach include assumptions about the degree of cloud cover, atmospheric ozone, photochemically inert natural organics and humic materials in the water, and ignoring light scattering. None the less, the technique offers rates computed from reproducible, readily obtainable data. Figure 6.7 illustrates a comparison of measured versus calculated half-lives for direct photolysis of a number of PAHs.

Indirect photolysis

The combined presence of humic substances, dissolved oxygen and solar radiation often results in an accelerated rate of organic chemical

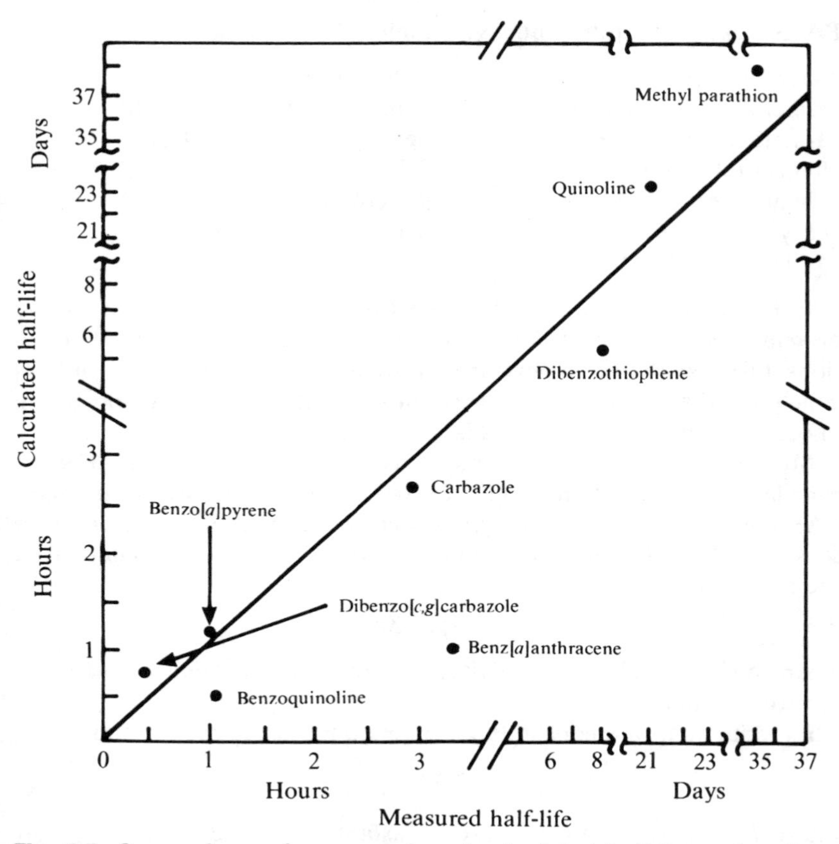

Fig. 6.7. Comparison of measured and calculated half-times for direct photolysis. Reproduced from Payne and Phillips (1985) (after Mill *et al.* (1981)).

transformation (Ross & Crosby, 1975; Zepp *et al.*, 1977*a*; Peyton & Glaze, 1988). Such sensitized photolysis involves the excitation of, say, a humic sensitizer by sunlight, followed by direct chemical interaction between the excited sensitizer and the chemical. The role of humic substances in photosensitized transformations in natural waters has been considered by Zepp *et al.* (1985). A second class of indirect photolysis involves the formation of chemical oxidants (Zepp & Baughman, 1978; Faust & Hoigné, 1987). The primary oxidants occurring in natural waters are hydroxyl and peroxy radicals (Mill *et* *al.*, 1980; Zepp *et al.*, 1987*a*; Kormann *et al.*, 1988), and singlet oxygen (Zepp *et al.*, 1977; Haag & Hoigné, 1986). Aquatic fate models describe the oxidative transformation of chemicals by a purely phenomenological coupling of a second-order rate constant to the concentration of oxidants in the system (Burns, 1982). Figure 6.8 illustrates three major pathways of known indirect photoreactions.

Almost all stable molecules are singlets (they have paired electrons in their ground states). The first transformation from the low-energy ground state to a higher energy electronically excited state by the absorption of a photon of ultraviolet light also results in a singlet. However, the energy transfer is normally too fast for spin inversion to occur and the singlet remains a short-lived species. Excited singlet states lose their excess electronic energy either by thermal reconversion to their ground state, fluorescence and a return to the singlet ground state, or by internal conversion (intersystem crossing) in which the singlet is transformed to a lower energy excited state with unpaired electrons (Larson & Berenbaum, 1988).

Molecules in triplet excited states normally have much longer

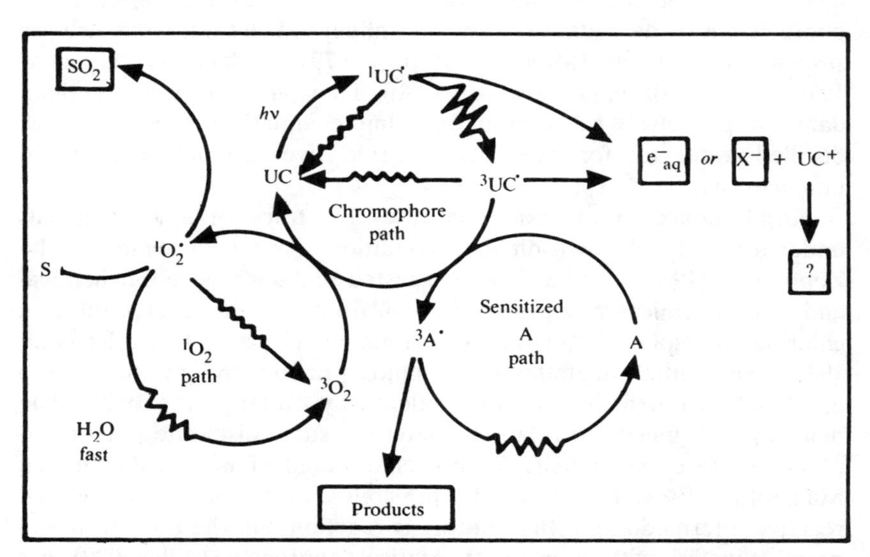

Fig. 6.8. Three major pathways for indirect photolysis. Reproduced from Zafiriou *et al.* (1984). UC, unknown chromophore; S, molecule reactive to singlet oxygen; A, energy acceptor other than oxygen; wavy arrows symbolize radiationless transitions.

lifetimes than singlets and are much more likely to engage in chemical reactions. Under certain conditions, the excited triplet (donor) can collide with a different molecule in the ground state (acceptor) and promote the acceptor to an excited state while itself returning to the ground state. Compounds capable of the efficient transfer of their triplet energy to donor molecules are known as photosensitizers. Those of possible environmental importance include the alkaloids berberine and quinine, butylated hydroxytoluene, bilirubin and related bile pigments, cercosporin and related fungal quinones, chlorophylls and degradation products, haems and related metalloporphyrins, humic substances, hypercin, metal oxide surfaces, naphthols, polyacetylenes and thiophenes, PAHs, protoporphyrins, riboflavin and other flavins, synthetic dyes and tetracyclines (Larson & Berenbaum, 1988).

Although qualitative considerations of potential photochemical reactivity may assume that the photoreactivity of molecules is independent of the nature of the aquatic medium (Harris, 1982*b*), quantitative prediction of photolysis rates requires more detailed investigation of the medium. Both the rate and the products of photochemical degradation can be influenced by factors such as suspended sediment (Miller & Zepp, 1979), surfactants (Hautala, 1978) and sensitizers (Zepp *et al.*, 1980). Thus, attempts to extrapolate data for photolysis rates from one simple aquatic medium such as distilled water to, for example, a turbid river are likely to prove unrewarding.

Nitrate-induced photooxidation rates for trace organic chemicals under a variety of environmental conditions have been estimated by Zepp *et al.* (1987*b*). It has been suggested that successive biochemical and photochemical processes may contribute to the degradation of chlorinated biphenyls in the environment (Baxter & Sutherland, 1984). Sequential degradation of chlorophenols by photolytic and microbial treatment has been considered by Miller *et al.*, 1988). The detection of linear correlations between such disparate factors as photolysis rate and toxicity for polychlorinated dibenzo-*p*-dioxins led Mamantov (1984) to suggest the possibility of a common or related reactive intermediate in the photolysis reaction and the biological end point. Finally, with a more theoretical approach, Mill (1989) has reviewed structure–activity relationships as a means of predicting environmental photooxidation rate constants.

6.4.2 Metals in water

Many data now exist on metal pollutants in water. However, most of these refer to total concentrations, thereby providing little information as to bioavailability and environmental mobility (Laxen, 1983*b*). In order to predict the environmental effects and fate of metals in water, it is necessary to determine physicochemical forms (species) and the interactions between them.

Figure 6.9 details possible metal forms in water, classified according to size. The simplest is the hydrated metal ion with a charge of +2 for many metal pollutants. Such ions may form loose electrostatic associations with anions to form ion pairs with a wide range of behaviour-determining charges. Stronger binding is demonstrated by organometallic compounds ranging in complexity from simple amino-acid complexes to those formed with humic substances, the latter being the predominant form of dissolved organic matter in natural waters. The kinetics of trace metal complexation has been considered with reference to the alkaline-earth metals by Hering and Morel (1988). A useful collection of data on organometallic compounds in the environment was provided by Craig (1983).

size μm			0·001	0·01 0·1	1
		soluble		colloidal	particulate
metal species	Free metal ions	Ion pairs; simple organic complexes; organo-metallic compounds	Complexes with high molecular weight organics	Metal species adsorbed on organic and inorganic colloids	Metal precipitates mineral solids; living and dead organisms
example	Cd^{2+}	$ZnHCO_3^+$ Cu-glycinate $(CH_3)_4Pb$	Cu-humics	Pb-FeOOH Co-MnOOH Cu-humics Hg-clay Cd-humic/FeOOH	$Cu_2S(s)$ $Pb_3(OH)_3(CO_3)_2(s)$ Zn-feldspar Ni-organic

Fig. 6.9. Possible metal forms in water classified according to size. Data from Laxen (1983*b*).

Metals existing as ions, ion pairs and simple low molecular weight compounds can be considered to be soluble or in true solution. Colloidal metal forms with a diameter of 0·001 to 1·0 μm and particulate forms of diameter greater than 1·0 μm are also common. Metals may be associated with particles that are inorganic, organic or mixtures of both.

Processes regulating interactions between different metal species include solubility control, adsorption, complexation, redox reactions methylation and biological uptake. Often solubility is regulated by the hydroxyl ion concentration and is thus a function of solution pH. Figure 6.10 illustrates theoretical metal hydroxide solubility as a

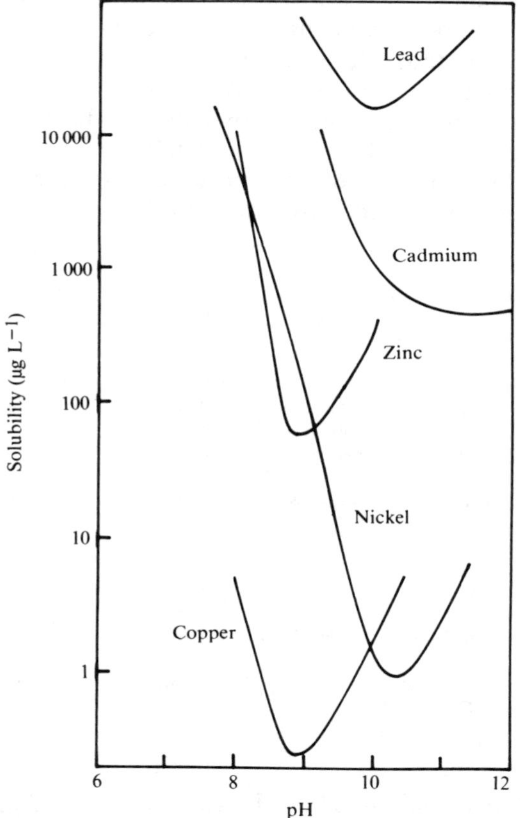

Fig. 6.10. Theoretical metal hydroxide solubility as a function of pH. Reproduced from Laxen (1983*b*).

function of pH. It is clear that metals differ widely in their response to such factors, therefore precluding the generalizations about the chemistry essential to simple models of pollutant environmental fate. The concentration of carbonate ions also significantly affects the solubility of metal salts. Aluminium chemistry, downstream of a whole-tree harvested watershed in the Hubbard Brook Ecosystem Study, has been described by Lawrence and Driscoll (1988).

Metals show considerable variability in their degree of association with organic substances. Such associations are generally held to be complexations, although metals may be adsorbed on to higher molecular weight humic substances (Laxen, 1983*b*). Mantoura *et al.* (1978) suggested that the order of complexation for a number of metals might be as follows:

$$Hg > Cu > Ni > Zn > Co > Mn > Cd$$

Even then, such rankings are only applicable under specified conditions. Variation of parameters such as pH may change metal species and so alter environmental behaviour and fate. Poisson and Papaud (1983) described experimental procedures for determining the individual contribution of five of the main constituents of seawater (Na^+, Ca^{2+}, Cl^-, SO_4^{2-} and HCO_3^-, to the transport of marine salt by diffusion. In this case, processes were influenced by salinity and temperature.

6.5 BIODEGRADATION

Biodegradation is one of the most important environmental processes in the breakdown of chemical compounds. It is a significant loss mechanism in soil (Kleopfer *et al.*, 1985) and aquatic systems and is essential to wastewater treatment (Scow, 1982). The eventual mineralization of organic compounds may be attributed almost entirely to biodegradation processes (Alexander, 1978). Although higher organisms can metabolize compounds (Varanasi *et al.*, 1985), microbial degradation plays the most significant role in environmental systems. The majority of biodegradation reactions can be categorized as oxidative, reductive, hydrolytic or conjugative (Hill, 1978). Table 6.4 presents examples of the first three kinds of reaction.

Abiotic and biotic transformations of halogenated aliphatic compounds have been reviewed by Vogel *et al.* (1987). The most likely

Table 6.4
Examples of biodegradation reactions

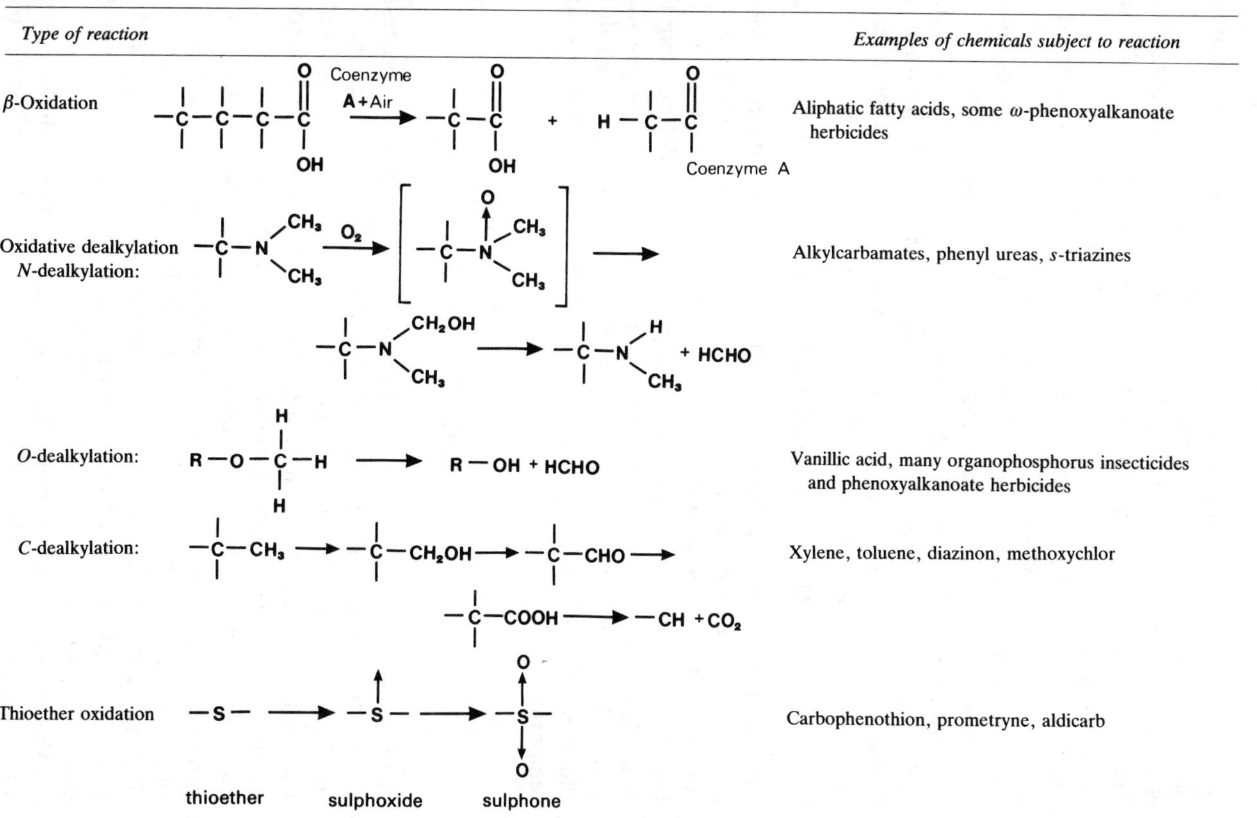

Type of reaction	Examples of chemicals subject to reaction
β-Oxidation	Aliphatic fatty acids, some ω-phenoxyalkanoate herbicides
Oxidative dealkylation *N*-dealkylation:	Alkylcarbamates, phenyl ureas, *s*-triazines
O-dealkylation:	Vanillic acid, many organophosphorus insecticides and phenoxyalkanoate herbicides
C-dealkylation:	Xylene, toluene, diazinon, methoxychlor
Thioether oxidation	Carbophenothion, prometryne, aldicarb

Decarboxylation	$-\overset{\vert}{\underset{\vert}{C}}-COOH \longrightarrow -\overset{\vert}{\underset{\vert}{C}}-H + CO_2$	Nicotinic acid, *o*-pyrocatechuic acid
Epoxidation	$-C{=}C- \xrightarrow{O_2}$ $-C(-O-)C-$	Aldrin, heptachlor
Aromatic hydroxylation		
(1) Aerobic:	Benzene $\xrightarrow{O_2 + 2H}$ Phenol (OH) $+ H_2O$	Pyridine, nicotinic acid, 2,4-D, some phenylalkanes, benzoic acid
	Benzene $\xrightarrow{O_2}$ Catechol (OH, OH)	
(2) Anaerobic:	Benzoic acid (COOH) $\xrightarrow{4H}$ Cyclohex-1-ene-1-carboxylate (COOH) $\xrightarrow{H_2O}$ 1-Hydroxycyclohexane-carboxylate (COOH, OH)	Benzoate

continued

Table 6.4—contd.

Type of reaction	*Examples of chemicals subject to reaction*
Aromatic, non-heterocyclic ring cleavage	
Ortho fission: OH, OH → (O_2) COOH, COOH	Many catechols and phenols, gentisic acid, hydroxycyclohexanecarboxylate, many phenoxy-alkanoate herbicides, carbaryl
Meta fission: OH, OH → (O_2) OH, COOH, CHO	Only first step in degradation pathway is shown.
'Gentisate' fission: HO, COOH, OH → (O_2) HO, COCOOH, COOH	
Aromatic, heterocyclic ring cleavage	
(1) 5-membered ring: indole → (O_2) 2,3-dihydroxy-indole (OH, OH, NH) → (O_2) [COOH, NHCOOH] → anthranilic acid (COOH, NH_2)	Many heterocyclic pesticides (e.g. paraquat, picloram, amitrole)

(2) 6-membered ring: Pyridines, pyrimidines, triazines

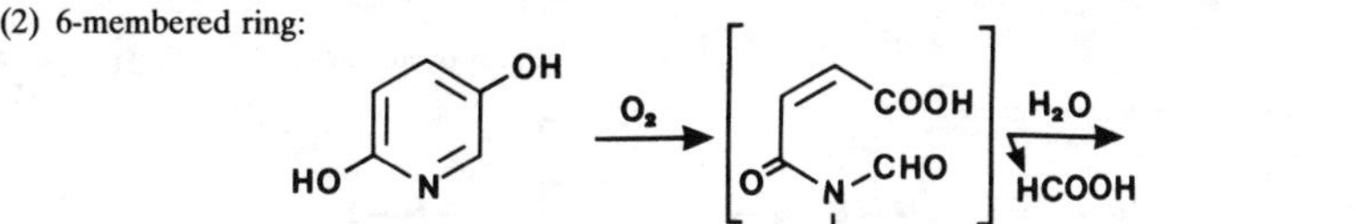

Hydrolysis

Ester hydrolysis: $R_1-C(=O)-O-R_2 \xrightarrow{H_2O} R_1-C(=O)-OH + HO-R_2$

Amide hydrolysis: $R_1-C(=O)-N(R_2)(R_3) \xrightarrow{H_2O} R_1-C(=O)-OH + H-N(R_2)(R_3)$

Phosphorus ester hydrolysis: $(R_1O)(R_2O)P(=O)-O-R_3 \xrightarrow{H_2O} R_1-OH + R_2-OH + H_3PO_4 + HO-R_3$

Nitrile hydrolysis: $R-C\equiv N \xrightarrow{H_2O} R-C(=O)-NH_2 \longrightarrow R-C(=O)-OH + NH_3$

Carbamates, organophosphates, many urea and anilide herbicides

continued

Table 6.4—contd.

Type of reaction		*Examples of chemicals subject to reaction*
Halogen reactions		
Hydrolytic dehalogenation:	$R-C(X)(H)-COOH \xrightarrow{H_2O} R-C(OH)(H)-COOH + X^-$ (aliphatic)	TCA, dalapon, halogenated phenoxyacetates, chlorobenzoates
	$C_6H_5X \xrightarrow{O_2} C_6H_5OH + X^-$ (aromatic)	
Halogen migration:	$R-C_6H_4-X$ (para) $\xrightarrow{O_2+2H}$ $R-C_6H_3(X)(OH)$ $+ H_2O$	Alkylbenzenes, tryptophan, halobenzenes, anisoles, 2,4-D
Reductive dehalogenation:	$-C-C(X)(X)- \xrightarrow{H_2} -C-C(H)(X)- + X^-$	*p,p'*-DDT, BHC
Dehydrohalogenation:	$-C(H)-C(X)(X)- \longrightarrow -C=C(X)- + HX$	*p,p'*-DDT, γ-BHC (lindane)
Nitro-reduction	$R-NO_2 \rightarrow [R-N=O \rightarrow R-N(OH)H] \rightarrow R-NH_2$	Parathion and other aromatic nitro-compounds

Data from Hill (1978). Reproduced from Scow (1982).

transformations to occur under given environmental conditions are controlled mainly by the number and type of halogen substituents. Increased halogenation or substitution of bromine for chlorine substituents increases the electrophilicity and oxidation state of the compound, making it more susceptible to dehydrohalogenation and reduction and less susceptible to substitution and oxidation. Oxidations and reductions are more common reactions in mammalian and microbiological systems, where they are mediated by enzymes or coenzymes. For oxidations, initial products are generally alcohols or epoxides. For reductions, products are generally less halogenated than their precursors. Potential mediators of such reactions are a variety of transition metal complexes, including iron porphyrins (Vogel *et al.*, 1987).

The interactions of metals and protons with algae have been studied by Crist *et al.* (1988), while a Weibull dose–response curve for the growth rate of algae as a function of the concentration of metallic and organic toxicants has been compared favourably with probit and logistic models (Nyholm, 1984).

In natural systems, among microorganisms predominantly responsible for biodegradation are the heterotrophic bacteria, including the actinomycetes, some autotrophic bacteria, fungi, including the basidiomycetes and yeasts, and some protozoa (Alexander, 1971). Microorganisms are found in diverse environments, although all will necessarily possess some aquatic component. The distribution of microbial populations in various aquatic systems is outlined in Table 6.5.

Variables potentially affecting the rate of biodegradation are either related to the nature of the compound, the nature of the organism or the type of environment. The physicochemical properties of the chemical and its environmental concentration can mediate between species composition, distribution and density. The previous history of the organism may be vital to its ability to biodegrade particular substances. Further complications may include inter- and intra-species interactions and enzymatic composition and activity. Typical environmental factors influencing biodegradation rates are temperature, pH, salinity, moisture and oxygen availability, the interactive effects of other substances and nutrient limitations (Lewis *et al.*, 1986; Swindoll *et al.*, 1988).

Several possible mechanisms, or combinations of mechanisms, may account for adaptation to xenobiotic compounds. Often time is

Table 6.5
Presence of microbial populations in various aquatic systems

	Upper layer of water	*Lower layer of water*	*Sediment layer*
Fresh water lotic (running water)	Microbial population highly dependent on stream flow, usually greater in slower streams and rivers		Presence of sediment layer and microbial population dependent on stream flow, surrounding substrate characteristics, and sediment load. Generally greater in slower flowing than in rapid streams
Lentic (standing water)	*Epilimnion:* Microbial populations primarily associated with this layer, although sediment populations greater under some conditions. Organisms associated with surface area of detritus. Fenchel and Jorgensen (1977)	*Hypolimnion:* Anaerobic microorganisms may be numerous in nutrient-rich eutrophic lakes	Microbial populations vary greatly, depending on depth, bottom substrate, and other factors. In most lakes, populations large near surface of sediment, although sediment investigations are rare. Gray (1978)
Estuarine	Biological activity greatest in lower bay (from river mouth upstream), in upper basins, and especially in tidal salt marshes and mud flats. Salinity shifts in the headwaters area of mixing of fresh and salt waters may be too extreme to support much life		Microbial populations large, especially in highly organic muds
Marine waters intertidal (littoral)	—	—	Microbial populations dependent on substrate: numerous on organic substrates, few on cobble and shingle beaches
Neritic	*Euphotic layer:* biological activity high. Organisms associated with surface detritus. Fenchel and Jorgensen (1977)	Most neritic waters fall within the euphotic zone	*Benthic layer:* microbial activity high
Oceanic	*Euphotic layer:* area of greatest microbial activity. Hobbie *et al.* (1972); Holm-Hansen (1972); Jannasch (1979)	*Aphotic layer:* microbial activity generally lower than in euphotic layer, although specific depths may have greater populations	*Benthic layer:* microbial activity reduced because of cold temperatures and low nutrient levels

Other sources: Odum (1971); Reid and Wood (1976).
Data from Scow (1982).

required for a degrader population to increase to the extent required to produce a measurable biodegradation rate (Spain *et al.*, 1980; Ventullo & Larson, 1986; Wiggins *et al.*, 1987) or genetic alteration (Wyndham, 1986). The influence of substrate concentrations and preexposure on the adaptation of aquifer microbial communities to xenobiotic compounds has been considered by Aelion *et al.* (1989).

Alexander (1985) warned against laboratory tests using chemical concentrations greater than those found in nature, since they may lead to erroneous inferences about microbial transformations. This is because there is a threshold below which certain, usually mineralizable molecules, are not converted to carbon dioxide. Thus, microorganisms may not assimilate carbon from chemicals present in trace amounts in natural environments, and so will not grow or produce the large acclimated populations needed for enhanced biodegradation. An automated gas-chromatographic measurement of evolved carbon dioxide has been suggested as a method of estimating biodegradability (Boatman *et al.*, 1986). Several microbial density enumeration techniques have been evaluated by Hickman & Novak (1989).

Some biodegradation rates may be estimated qualitatively by rules-of-thumb based on structural factors and chemical class (Table 6.6). Alternatively, correlations have been made between the biodegradability of certain chemical groups and fundamental properties. However, in general, inadequate databases prevent quantification of such relationships. Further details on the estimation of biodegradation rates were given by Scow (1982).

In a study commissioned by the UK and Malaysian divisions of Shell Petroleum Company, the biodegradation of crude oil in the marine environment was evaluated for Sabah and Sarawak (UKM/USM/SRC, 1986). This comprehensive report considered the distribution of hydrocarbon-utilizing microorganisms, determined hydrocarbon biodegradation potentials and measured rates of biodegradation and photooxidation. Field validation of laboratory studies highlighted the importance of photooxidation on Baram crude, which was particularly sensitive to sunlight, releasing organic carbon into the water column to be degraded with time.

Photolysis was the primary transformation process for polychlorinated phenols with photolysis rate constants in surface estuarine water ranging from 0·3 to $1{\cdot}2\,h^{-1}$ and half-lives ranging from 0·6 to 3 h. Photoproducts were then rapidly degraded by microbes (Hwang *et al.*, 1986). Microbial degradation was the primary process for transformation

Table 6.6
Rules of thumb for biodegradability

Factors	*Schematic example*[a]
Branching (highly branched compounds are more resistant to biodegradation)	*Branching*
(1) Unbranched side chains on phenolic and phenoxy compounds are more easily metabolized than branch alkyl moieties (Webley *et al.*, 1959)	$CH_3-CH_2-CH_2-CH_2-CH_3$ > $(CH_3)_2CH-CH_2-C(CH_3)(CH_2CH_3)-CH_3$
(2) 2,4-Dichlorophenoxyalkanates with side chains of four or more carbons degraded easily, the propionate more slowly, and dichlorophenoxyacetate not at all by a *Flavobacterium* sp. (MacRae & Alexander, 1963)	$HO-C_6H_4-CH_2CH_2CH_3$ > $HO-C_6H_4-CH(CH_3)_2$
(3) Branched alkyl benzene sulphonates degrade more slowly than straight chain (Swisher, 1963*a*).	
Chain length (short chains are not as quickly degraded as long chains)	*Chain length*
(1) Rate of oxidation of straight-chain aliphatic hydrocarbons is correlated to length of chain (Ladd, 1956)	$CH_3-CH_2-CH_2-CH_2-CH_2-CH_2-CH_3$ > $CH_3-CH_2-CH_3$
(2) Soil micros attack long-chain mononuclear aromatics faster than short-chain (Strawinski, 1943)	
(3) Micros grow on normal alkanes from *n*-octane to *n*-eicosane but not on *n*-heptane to methane (Finnerty *et al.*, 1962)	
(4) Sulphate-reducing bacteria more rapidly degrade long-length carbon chains (decane to hentriacontane) than short-length carbon chains (Bokova, 1954)	
(5) ABS detergents increase in degradability with increase in chain length from C_6 to C_{12} but not $>C_{12}$ (Huddleston & Allred 1963; Swisher, 1963*a*)	
(6) Rate of mineralization of N in urea–formaldehyde complexes declines with increasing ureaform chain (Long & Winsor, 1960)	

Oxidation Highly oxidized compounds like halogenated compounds may resist further oxidation under aerobic conditions but may be more rapidly degraded under anaerobic conditions (Hill & McCarty, 1967; Helling *et al.*, 1971, Hamaker, 1972*a*; Goring *et al.*, 1975)	
Polarity and ionization Nonionic compounds with active halogens are likely to be degraded by nucleophilic displacement reactions like hydrolysis. The same is true of linkages separating highly polar groups (Goring *et al.*, 1975)	*Polarity* $CH_3-C(=O)-O-NO_2$ > $CH_3-C(=O)-O-CH_3$
Saturation Unsaturated aliphatics are more readily attacked than corresponding saturated hydrocarbons, perhaps because of presence of many ethylene-reducing enzyme systems and few ethane ones	*Saturation* $CH_2=CH_2$ > CH_3-CH_3
Substituents (*number of*) *on simple organic molecules*	*Substituents* (*number of*)
(1) Alcohols, aldehydes, acids, esters, amides and amino acids are more susceptible to biodegradation than the corresponding alkanes, olefins, ketones, dicarboxylic acids, nitriles, amines and chloroalkanes (Painter, 1974)	$CH_2OH-CH_2-CH_3$ > $CH_3-CH_2-CH_3$ NH_2-CH_2-COOH > NH_2-CH_3 $CH_3-CH(=O)$ > $CH_3-C(=O)-CH_3$
(2) Increased substitution hinders oxidation responsible for breakdown of alkyl chains (Hammond & Alexander, 1972)	$CH_3-CH_2-CH_2-CH_2-R$ > $CH_3-CH(OH)-CH(Cl)-CH(NH_2)-R$
(3) No significant oxidation of polycyclic aromatic hydrocarbons containing more than three rings (McKenna & Heath, 1976)	>
(4) Longer persistence of chloroacetic acids, α-substituted propionic acids, isopropyl *N*-phenylcarbamates, and isopropyl phenylcarbamates with greater number of chlorines (Kaufman, 1966)	

continued

Table 6.6—contd.

Factors	*Schematic example*[a]
(5) Diaminobenzenes show less availability than monaminobenzenes (Alexander & Lustigman, 1966)	
(6) On aromatic ring: benzoic acid quickly degrades; monochloro- and monofluoro-benzoates are more resistant but can be degraded; di-, tri- and tetra- are quite resistant. The more chlorines, the more resistant the compound (MacRae & Alexander, 1963)	COOH > $C(=O)-O^{-}-Na^{+}$, Cl > $C(=O)-O^{-}-Na^{+}$, Cl, Cl
(7) Presence of more than one methyl group attached to a carbon strongly inhibits alkane utilization (Manahan, 1969)	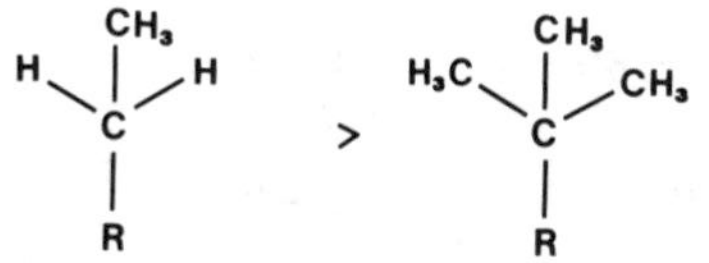 H, CH_3, H, C, R > H_3C, CH_3, CH_3, C, R
Substituents (position of) on simple organic molecules	*Substituents (position)*
(1) Aromatics with methyl, chloro, nitro, or amino are generally not available to bacteria, which are active on unsubstituted molecules, or they are decomposed more slowly; *para* is more utilized than *ortho* or *meta*. (Kramer & Doetsch, 1950; Bell, 1960; Ali *et al.* 1962; MacRae & Alexander, 1963; Chambers & Kabler, 1964; Tabak *et al.*, 1964; Henderson, 1975)	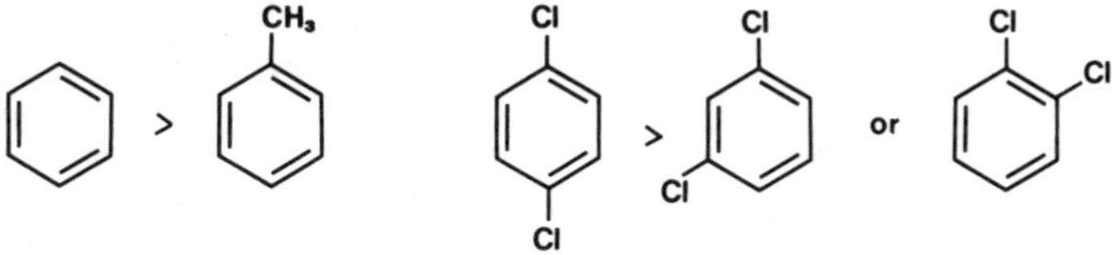 > CH_3 Cl, Cl > Cl, Cl or Cl, Cl
(2) *Meta*-disubstituted phenols and phenoxys more resistant than *ortho* or *para* isomers (Alexander & Aleem, 1961)	OH, Cl or OH, Cl > OH, Cl
(3) Mono-, di- and tri-chlorophenols with halogen *meta* to hydroxyl, phenoxy-alkanoic acids with chlorine *meta* to ether oxygen, and benzoic acids with *meta*-amino, nitro, or methoxy groups are less readily degraded than corresponding isomers with substituents in *ortho* or *para* positions (Alexander & Aleem, 1961; Burger *et al.*, 1962; MacRae & Alexander, 1963; Alexander & Lustigman, 1966; Kearney & Plimmer, 1970)	

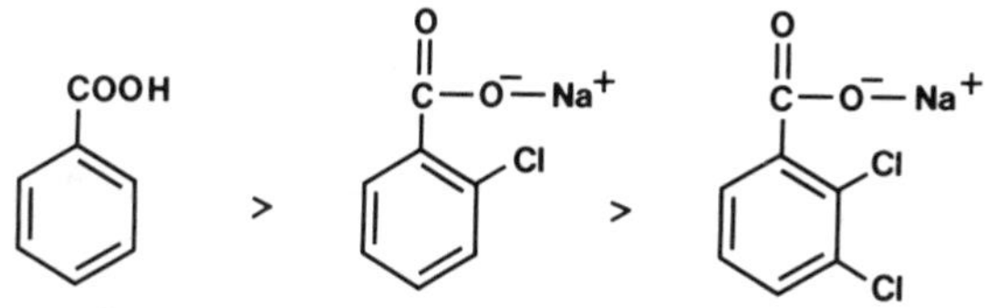

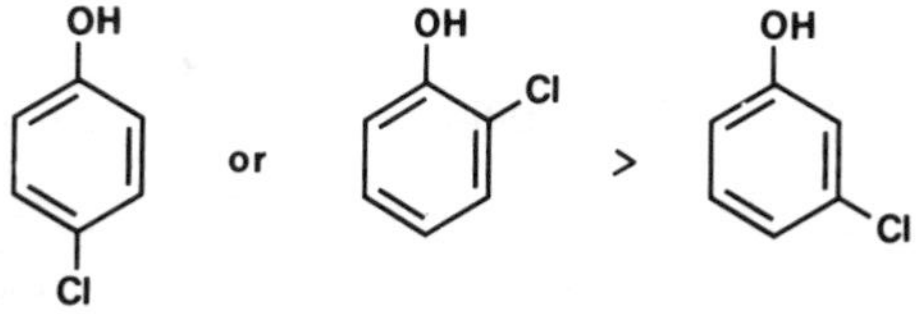

(4) Conversely, *ortho* isomers of nitrophenols, methylanilines, sulphonates of 1-phenyldodecane, and chlorine-containing isopropyl phenylcarbonates are most persistent (Swisher, 1963*b*; Alexander & Lustigman, 1966; Kaufman, 1966)	*m*-nitrophenol (OH, NO_2) or *p*-nitrophenol (OH, NO_2) > *o*-nitrophenol (OH, NO_2)
(5) In fatty acids, introduction of halogen or phenyl group on α-carbon reduces rate of degradation as opposed to same group on omega carbon (Burger *et al.*, 1962; Dias & Alexander, 1971)	C_6H_5–$CH_2(CH_2)_5COOH$ > $CH_3(CH_2)_4CH(C_6H_5)COOH$
(6) *p*-Hydroxybenzoate degrades more rapidly than *ortho* or *meta* (fewer micros degrade these) (Stanier, 1948)	
(7) For ABS, *para*-sulphonates are more readily degraded than *ortho*-sulphonates of phenyldodecane and phenyltetradecane. In diheptylbenzene sulphonates the *meta* is more-susceptible than the *para* substituent (Swisher, 1963*a,b*)	
(8) Neopentyl group additon inhibits alkane utilization if carbon atom bonded is next to last on the chain (Manahan, 1969)	$CH_3CH_2CH(CH_2C(CH_3)_2CH_3)CH_2CH_2$—R > $CH_3CH(CH_2C(CH_3)_2CH_3)CH_2CH_2CH_2$—R
Substituents (*type of*) *on simple organic molecules*	*Substituent* (*type*)
(1) Mono- and di-carboxylic acids, aliphatic alcohols and ABS are decreasingly degraded when hydrogen is replaced by methyl groups (Swisher, 1970; Dias & Alexander, 1971; Hammond & Alexander, 1972)	CH_3—CH_2—COOH > CH_3—$C(CH_3)_2$—COOH
(2) Aliphatic acids are less easily degraded when chlorine replaces a hydrogen (Dias & Alexander, 1971)	Cl—CH(Cl)—C(=O)—OH > Cl—C(Cl)(Cl)—C(=O)—OH
(3) Triazines and methoxychlor are less easily degraded when methoxy groups are replaced by chlorines (Hauck & Stephenson, 1964)	CH_3—O—C_6H_4—CH(CCl_3)—C_6H_4—O—CH_3 > Cl—C_6H_4—CH(CCl_3)—C_6H_4—O—CH_3

continued

Table 6.6—contd.

Factors	Schematic example[a]
(4) Degradation of disubstituted benzenes is reduced when carbonyl or hydroxyl is replaced by nitro, sulphonate or chloro group (Alexander & Lustigman, 1966)	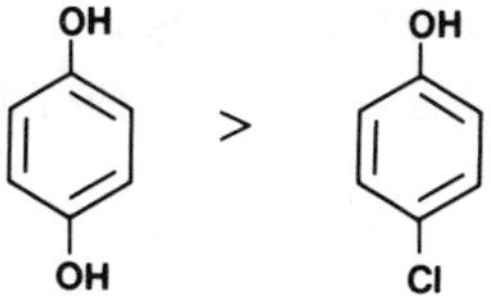 OH / OH > OH / Cl
(5) Successive replacement of hydroxyls of cyanuric acid with amino groups makes compounds less degradable (Huack & Stephenson, 1964)	
(6) For naphthalene compounds, nuclei-bearing, single, small alkyl groups (methyl, ethyl or vinyl) oxidize more rapidly than those with a phenyl substituent (McKenna & Heath, 1976)	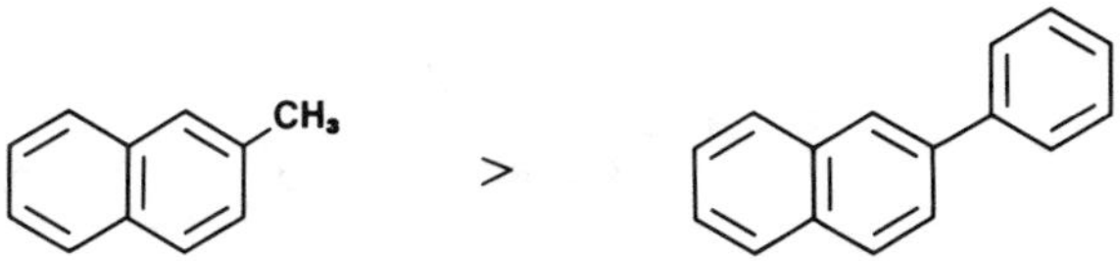 CH_3 >
(7) 3-Phenyl-1,1-dimethylurea (fenuron) is more rapidly degraded than 3-(*p*-chlorophenyl) · · · (monuron), or 3-(3,4-dichlorophenyl) · · · (diuron) (Sheets, 1958)	

(8) *N*-substituent methyl and ethyl groups in anilines are harder to oxidize (Alexander, 1965)

(9) Dehalogenation may be specific for the halide substituent (e.g. -chloropropionate, not 3-bromo-; tribromoacetate, not trifluoro-) (Hirsch & Alexander, 1960)

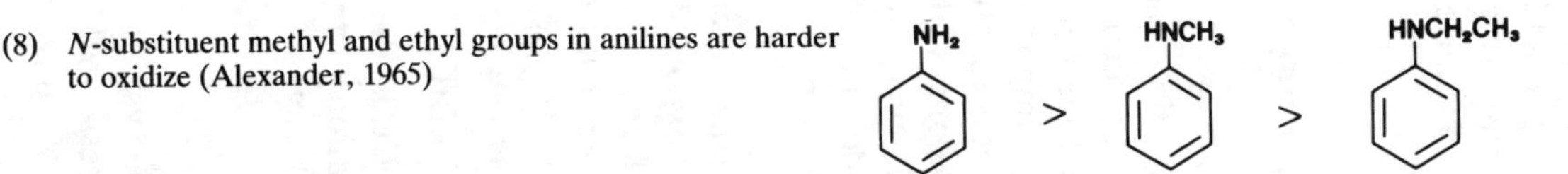

Miscellaneous

(1) Ether functions are sometimes particularly resistant to biodegradation (Sawyer & Ryckman, 1957)

(2) Alkyl and quaternary groups attached via oxygen to phosphorus atoms increase stability, possibly through steric blocking of ester bond (Nesbitt & Watson, 1980)

---

[a] '>' implies 'more easily degradable than'.
Adapted from Scow (1982).

of phenol and *p*-chlorophenol. Banerjee *et al.* (1984) developed a general kinetic model for biodegradation and applied it to chlorophenols and related compounds. They found that, in most cases, rate decreased with increasing lipophilicity of the substrate. Subsequent published correspondence argued the case for and against, respectively, enzymatic processes rather than transport phenomena as an interpretation of experimental findings (Banerjee *et al.*, 1985; Beltrame *et al.*, 1985).

6.6 MATHEMATICAL MODELS

6.6.1 Environmental rates approach

Once data are expressed as time–concentration rates, they can be incorporated into a suitable model for predicting environmental concentrations or ranking chemicals relative to a 'benchmark' chemical (Branson, 1978). The simplest and yet most important principle used in constructing such models is the law of conservation of matter. In real cases, accounting boundaries may accord with actual physical discontinuities such as river banks or the air–water interface. In many cases, the accuracy and detail of a model may be improved by subdivision into arbitrary lines drawn on a map or vertical profile of the water body.

Mass balance demands that every molecule entering the environmental phase under consideration is either exported, transformed or retained. The mass balance can therefore express the change in concentration. Given that the accountable mass is the product of concentration (mass/volume) and volume, the rate of change of concentration $[C]$ per unit time (mass change rate) is as follows:

$$V\,\mathrm{d}[C]/\mathrm{d}t = Lo - Ex - VkE[C] \qquad (9)$$

where V is the volume, Lo is the load, Ex is the export, E is an environmental factor driving the transformation of the compound and k is the kinetic constant of the reaction pathway.

For water-borne exports only:

$$Ex = F[C] \qquad (10)$$

where F is the flow rate of water leaving the system (volume/time).

The second-order transformation process $kE[C]$ has units of mass/volume/time, but multiplication by V converts it to units of mass/time (Burns, 1983).

Dividing both sides of eqn (9) by V gives a unit equation for modelling chemical concentrations in real systems as follows:

$$d[C]/dt = Lo/V - (F/V)[C] - kE[C] \qquad (11)$$

where (V/F) is the 'hydraulic residence time'.

Although export processes may often be more complicated than the expression given in the above equation (e.g. volatilization across the air–water interface) the majority of environmental fate models are in essence more complex variations of the mass-balance themes in eqn (11).

Kenaga (1968) identified many of the laboratory data necessary to assess the fate of pesticides in the environment, using the environmental rates approach. In theory, the number of properties that might improve the accuracy of a prediction are not limited, but in practice there are likely to be diminishing returns as the data set increases. If approximate calculations based on a limited data set yield adequate answers, then it is not necessary to refine the model further.

Several typical properties of chemicals used to make predictions of environmental concentration are listed in Table 6.7. Some of the properties are sufficiently fundamental and general in nature to be available early in the modelling process. The required rate and equilibrium constants can be determined from the fundamental properties of the chemical and simple laboratory screening tests (Johnson & Parrish, 1978). It should be emphasized that not all the properties listed will be necessary for every chemical.

The typical properties of the environment used to make predictions of aquatic fate and concentration are given in Table 6.8. Here, the property selected will be determined by the nature of the specific environmental compartment under consideration. As before, not all the properties cited in Table 6.8 will necessarily be required for the prediction of a chemical's environmental fate.

Using a pond model and laboratory data, Neely and Blau (1977) used the materials balance equation outlined in Fig. 6.11 for predicting the fate of chloropyrifos. There was good agreement between predicted and experimentally found concentrations in the fish and water, using this model. The factor k_0 (the rate of input), in Fig. 6.11, is an essential parameter, which must adequately reflect the influence of the chemical.

The environmental rates approach is potentially useful in ranking chemicals in terms of hazard to aquatic organisms based on calculated

Table 6.7
Properties of a chemical useful in predicting the concentration of the chemical in aquatic environments[a]

Property	
Molecular structure	
Water solubility	
Vapour pressure	
Absorption spectra (ultraviolet, visible)	
Particle size (if substance is particulate)	
Rate constants	*Partition coefficients*
Photodegradation (ultraviolet, visible)	octanol : water
Biological degradation	air : water
Chemical degradation	sediment : water
Evaporation	
Sediment binding	
Uptake by organisms	
Depuration by organisms	

[a] Methods for deriving these properties were prepared by the Task Groups of ASTM Subcommittee E35.21 on Safety to Man and Environment.
Data from Johnson and Parrish (1978).

Table 6.8
Properties of aquatic environments useful in predicting the fate and concentration of a chemical in those environments

Property
Surface area
Depth
pH
Flow/turbulence
Carbon in sediment
Temperature
Salinity
Suspended sediment concentration
Trophic status
Absorption spectra (ultraviolet, visible)

Data from Johnson and Parrish (1978).

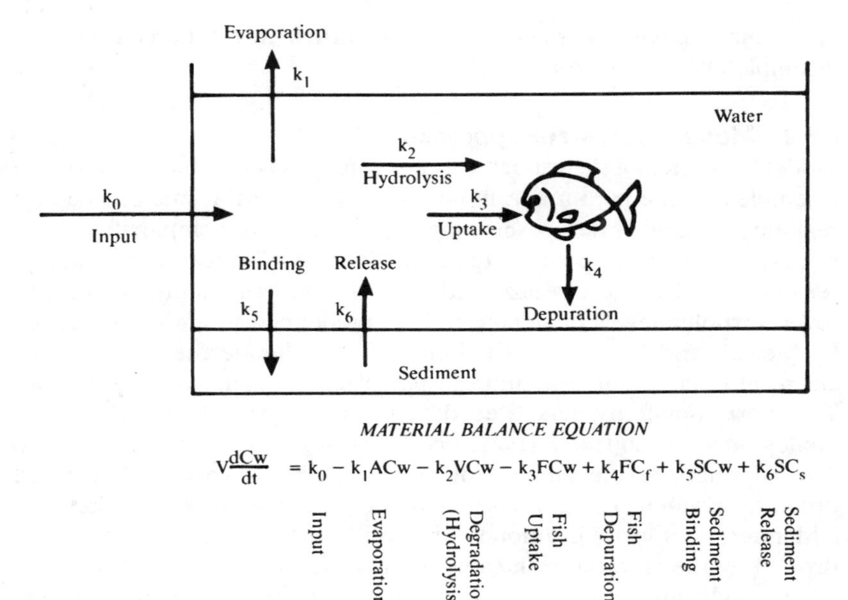

Fig. 6.11. The Neely and Blau (1977) Pond Model. Reproduced from Branson (1978). V, Volume of water; A, Surface area, cm^2; F, Fish mass, gm; S, Sediment mass, gm; Cw, Concentration of chemical in water; k, Rate constant; C_f, Concentration of chemical in fish; C_s, Concentration of chemical in sediment.

residue concentrations in tissues. The relative importance of each property, chemical or environmental, may be isolated and quantified by the 'sensitivity analysis' described by Branson (1978). This is achieved by hypothetically varying properties in turn, in order to establish which has the greatest effect on the others. Other advantages of the environmental rates approach are that the probable environmental fate and concentration of many chemicals can be rapidly estimated with minimum data. Also, rate data allow for an understanding of aquatic fate and concentration over a period of time. The greatest limitation to the environmental rates approach is the paucity of material balance data from field studies to validate the mathematical concepts involved. Other problems are that a single model is unlikely to be suitable for every chemical or environment. Furthermore, many environmental processes and their interactions are as yet incompletely understood.

6.6.2 Model ecosystem approach

Model ecosystem fate studies are appealing because they purport to resemble a real environment in miniature or at least a small piece of a real environment. Microcosms may vary in size and complexity from a bacterial isolate in culture to a small lake. Typical experimental set-ups are of aquarium size and contain soil, sediments, water and several trophic levels of organism. The model ecosystem as developed by Metcalf and Lu (1973) was intended to estimate the potential for chemical bioaccumulation and the potential for chemical degradation. The most commonly generated data from the Metcalf model are the biodegradability index (BI) and ecological magnification (EM) values. The BI refers to the amount of polar metabolite extracted from an organism divided by the relative amount of nonpolar metabolite. The EM refers to the total amount of parent compound in the organism divided by the amount in the surrounding water.

Although microcosms do not allow for prediction of absolute concentrations of a chemical expected or observed in all environments, standardization of microcosms allows comparison of different chemicals. The compartmentalization and fate of lindane, naphthalene and mirex were assessed in just such a standardized system by Saleh *et al.* (1982).

The National Research Council of Canada (NRCC, 1981) developed an empirical model as a screen for fundamental indicators of general chemical behaviour patterns. The relatively simple screening protocol adopted did not attempt to generate absolute values or simulate natural systems. Its role was to facilitate the identification of appropriate 'benchmarks' (e.g. chemicals) whose overall fate and persistence patterns were similar to those of environmental pollutants. Some characteristics of ideal benchmarks were as follows:

A. In terms of chemical reactivity, the compounds chosen must represent the various types of man-made industrial, agricultural and other organic chemicals in use. In other words, the compound must be susceptible to processes such as volatilization, hydrolysis, photolysis and biodegradation.

B. Compounds with different sorptive properties and with high and low octanol–water partition coefficients must be included within each class of chemical reactivity represented. Consider two hypothetical compounds that are readily degraded only by anaerobic bacteria. If one compound is highly hydrophobic and the other very soluble in water, then the soluble one will tend to partition into the water compartment while the other will partition into the sediment. Since anaerobic microbial degradation is usually most important in the sediment compartment, this process will be much more important for the hydrophobic compounds and, consequently, other things being equal, it will be less persistent than the soluble compound.

The NRCC generalized their view of an aquatic environment to the minimum number of compartments that still provide some flexibility

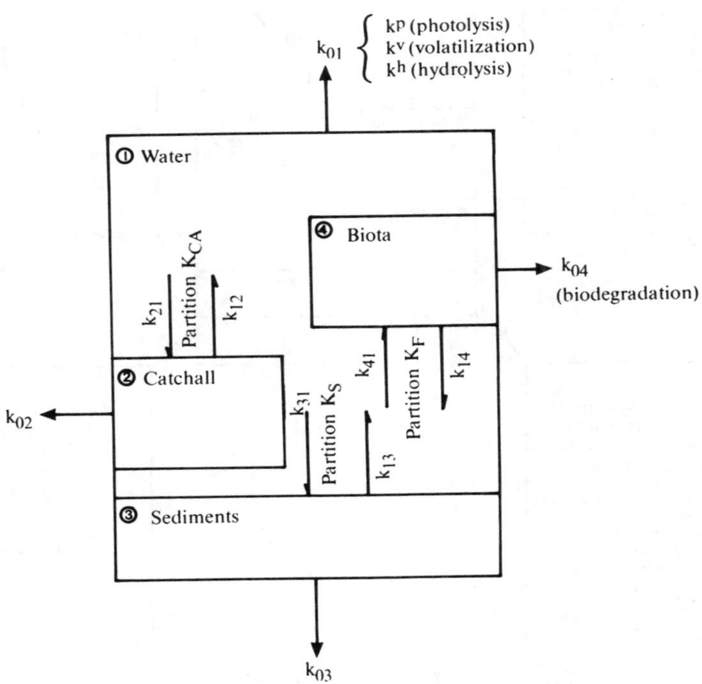

Fig. 6.12. Schematic of the four-compartment model. k_{12}, k_{13}, k_{14}, k_{21}, k_{31}, k_{41} = first-order rate constants; K_{CA}, K_S, K_F = partition coefficients between each compartment and water; k_{01}, k_{02}, k_{03}, k_{04} = first-order removal rate constants. Reproduced from NRCC, No. 18570 (1981).

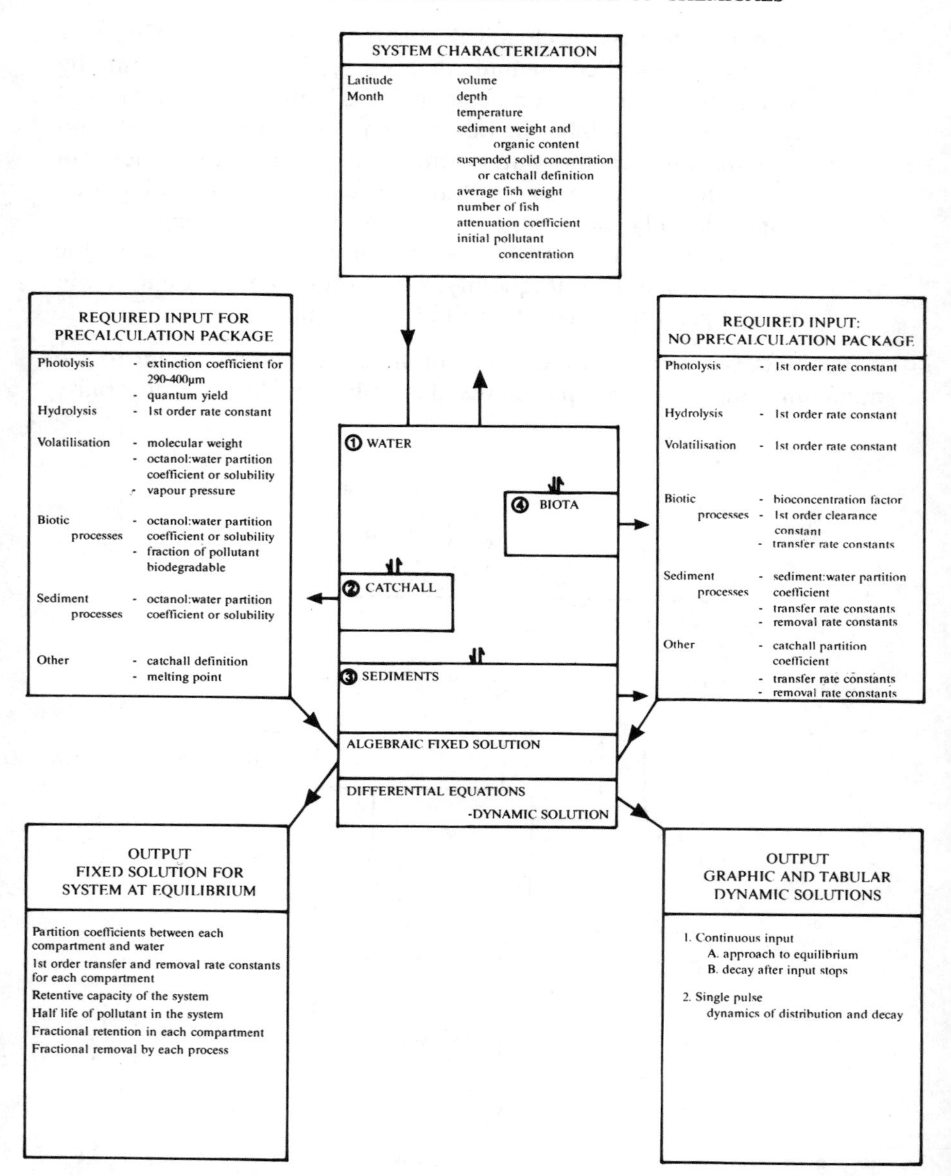

Fig. 6.13. Detailed schematic of the computer program. Reproduced from NRCC, No. 18570 (1981).

and realism. They chose to examine a four-compartment model (Figs 6.12 and 6.13) in which the water, sediment and fish compartments were clearly identified. A 'catch all' compartment was included to allow consideration of other potentially significant sorptive compartments such as aquatic plants or suspended solids. Constraints and assumptions to the model were:

(a) Only well-characterized chemical and physical processes were included in this analysis of pollutant fate.
(b) Hydrodynamic properties (flow rates, etc.) and settling effects of aquatic systems were excluded.
(c) The size of each compartment remained constant and the rate of mixing of the pollutant was rapid, relative to the rates of all other processes, thus allowing each compartment to be treated as homogeneous.
(d) Pollutant concentrations were always below their solubility limits. Therefore, a pure chemical phase did not exist. All processes were assumed to follow simple kinetics and to be approximated to by a first-order or pseudo first-order rate constant.
(e) An equilibrium would eventually describe the distribution of a chemical between the compartments in the system.

This screening protocol allowed for an estimate to be made of chemical persistence in terms of the inherent capacity of individual water bodies to retain a chemical. This property has been termed the *retentive capacity* of the water body (Roberts & Marshall, 1980) and is in effect the mean residence time of a particular chemical in an ecosystem.

6.6.3 Aquatic fate models

A wide range of aquatic fate models have been developed by organizations such as the US Environmental Protection Agency. Such models are collated in Table 6.9 along with summary details of their main features. The models chosen were at a reasonably advanced stage of developement and had existing computer software packages. No particular model is recommended, as a choice must depend on the particular chemical, environment and envisaged utilization. Certainly some models are more appropriate to particular modelling strategies.

For example, the EXAMS model was developed to incorporate the concept of fugacity into the rapid evaluation of synthetic organic

Table 6.9
Aquatic fate models

Title	*Model details*	*Reference*
SERATRA	Sediment Radionuclide Transport Model —based on a summary of process models published by Stanford Research Institute, Battelle Pacific NW Laboratories' SERATRA includes process models for photolysis, hydrolysis, oxidation, biolysis and volatilization —unsteady, two-dimensional (longitudinal and vertical) finite element model for transport of dissolved constituents and transport, deposition and resuspension of sediments and sorbed contaminants	Smith *et al.* (1977) Onishi and Wise (1982)
TODAM	—one-dimensional (longitudinal) version of SERATRA	
FETRA	—two-dimensional (longitudinal and lateral) estuarine version of SERATRA—inputs are timed hydrodynamic conditions. —sorption/desorption are treated as first-order kinetic processes in chemical concentrations	
CMRA	Chemical Migration and Risk Assessment —developed for use with the SERATRA program —includes a non point-source loading model for agricultural sources and a hydrodynamic model —also includes a toxicological post-processor FRANCO (FRequency ANalysis of pesticide COncentrations for risk assessment)	Onishi *et al.* (1982)
HSPF	Hydrologic Simulation Program —based on the Stanford Watershed Model —includes the generation of transformation products, each of which is in turn subject to volatilization, phototransformation, biolysis, etc.	Johanson *et al.* (1980)

HSPF 7.	—incorporates the process models of SERATA in its aquatic sections, with several user-selectable options for sediment transport	
WASP	Water Quality Analysis Simulation Program —generalized finite-difference model designed to accept user-specified kinetic models as subroutines —can be applied to one-, two- or three-dimensional descriptions of water bodies. —process models can be structured to include linear and nonlinear kinetics.	Di Toro *et al.* (1981*a*)
TOXIWASP/ WASTOX	—Two versions of WASP designed specifically for synthetic organic chemicals —both treat sorption–desorption as local equilibria —both include process models for biolysis, oxidation, volatilization and photolysis —These programs allow for the selection of either constant or time-variable transport and reaction processes	Ambrose *et al.* (1983) Connolly (1982)
TOXIC	—developed at the University of Iowa as an elaboration of the SRI model —to its predecessor, it adds a fish uptake and depuration model —has expanded dynamic capabilities	Schnoor *et al.* (1983) Schnoor (1981) Schnoor and McAvoy (1981)
UTM-TOX	Unified Transport Model for toxicants —based on the Oak Ridge National Laboratory UTM model —includes air, water and terrestrial submodels —aquatic fate submodel includes volatilization, hydrolysis, photolysis, biolysis and sorption equilibria —sorbed phases are assumed to be unreactive in the 1982 version	Browman *et al.* (1982)
MEXAMS	Metals Exposure Analysis Modeling System —links MINTEQ, a thermodynamic equilibrium model with EXAMS, an aquatic exposure assessment model	Felmy *et al.* (1984)

pollutants. Given the characteristics of a pollutant and an aquatic system, EXAMS computes steady-rate distributions of pollutant concentrations, the fate of the pollutant in the system, and the time required for effective purification of the system (persistence). The Metals Exposure Analysis Modeling System (MEXAMS), links MINTEQ, a geochemical model, with EXAMS. MINTEQ is a thermodynamic equilibrium model capable of computing aqueous speciation, adsorption and precipitation or dissolution of solid phases (Felmy *et al.*, 1984).

Chapter 7

The Atmospheric Environment

7.1 INTRODUCTION

The atmosphere may be divided roughly into three zones. The uppermost is the stratosphere, which extends to a height of some 50 km above the surface from about 8 km above the poles and about 16 km above the equator, subject to seasonal and other fluctuations. It is bounded by temperature inversions and is characterized by rising temperature with increasing height. The pressure at the lower bound is of the order of one-tenth that at the surface, and at the upper bound about one-thousandth of that pressure. It is generally calm because of its temperature gradient, which limits convection, and is free of cloud except near its base and free of precipitation. Regions above the stratosphere do not appear as yet to have produced issues important for pollution or environmental fate modelling studies (Sugden, 1979*a*).

Below the stratosphere the troposphere extends downwards almost to the Earth's surface with a temperature gradient in the opposite direction. The upper interface, the tropopause, is not readily crossed by diffusing gases. Our understanding of transfer across this interface is incomplete, especially in middle latitudes where it is ill defined, being spread over an extent of a few kilometres called the tropopause gap in which the temperature gradient is indeterminate (Sugden, 1979*a*). In terms of motion, the troposphere is relatively violent, since it responds vigorously to the uneven heating of land masses and oceans with the production of winds. It is associated with cloudiness, with the precipitation of rain or snow, and is very important in determining general patterns of climate.

Close to the surface of the Earth the troposphere is modified in a very roughly defined boundary layer, extending vertically for about a kilometre. In this zone, the atmosphere is particularly responsive to detailed features of the surface such as mountains, valleys, coastal littorals, conurbations and even large individual buildings. At times its upper bound may be associated with a temperature inversion, providing a region from which the upward escape of air is difficult. Figure 7.1 illustrates the general features of the atmosphere and provides details of the temperature profile and variation in density with altitude that occurs. A useful account of the physics of planetary atmospheres may be found in Houghton (1977).

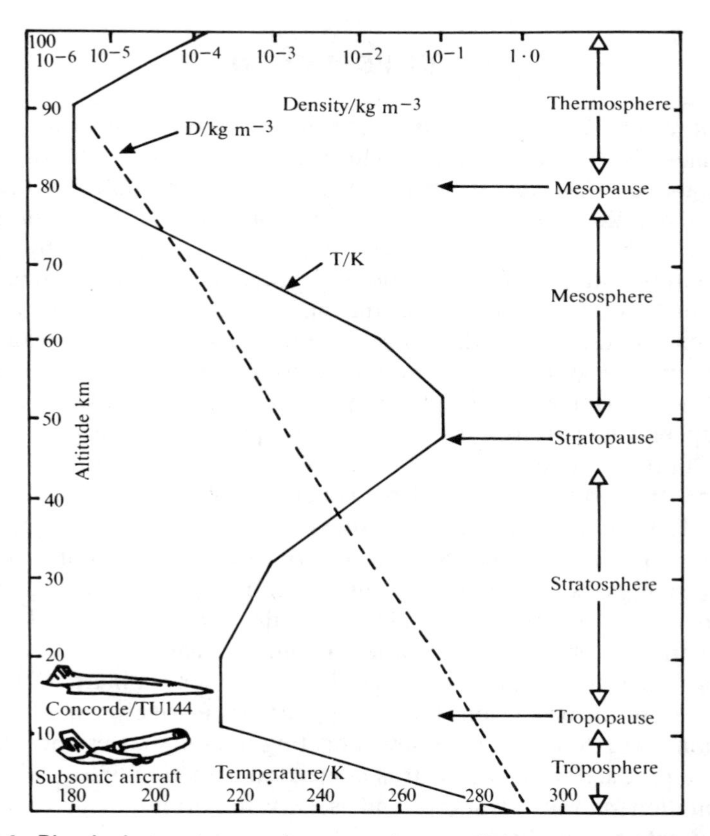

Fig. 7.1. Physical structure and nomenclature of the atmosphere. Reproduced from Donovan (1983).

The ion balance in the atmosphere is determined by relatively few chemical species. Over land, the main contributors of acidity are, in order of importance, SO_4^{2-} and NO_3^-, while over the oceans they are Cl^- and SO_4^{2-}. Contributors to alkalinity over land are Ca^{2+}, Mg^{2+}, NH_4^+ and K^+, whereas over the oceans the major species order is Na^+, Mg^{2+}, Ca^{2+} and K^+ (Summers, 1983).

Xenobiotic compounds encountered in the atmosphere are likely to possess a high degree of chemical stability, appreciable vapour pressure ($>10^{-7}$ mm Hg), a history of widespread use as foliar sprays and be of inherent high analytical detectability (Seiber *et al.*, 1983). Such conditions are, for example, met by the organochlorine compounds. The distribution of chemicals between vapour, liquid aerosol and solid particulate phases is dynamic, being determined by factors such as vapour pressure and concentration and by the size, surface area and organic content of the particulate load (Junge, 1977).

With the atmospheric compartment, the distinction between physical processes such as wind dispersal and chemical processes such as oxidation and photooxidation is not resolved into separate environmental fate models. Rather, the more complex atmospheric dispersion models incorporate the influence of chemical factors. Within this chapter are outlined some major atmospheric acidification pathways, with reference to sulphur dioxide, the oxides of nitrogen and the chlorofluorocarbons, as illustration of some of the physicochemical processes determining chemical fate in the atmosphere. Processes removing chemicals from the atmosphere are discussed in Section 7.3, which is followed by a discussion of atmospheric residence time. Existing atmospheric fate models are described in Section 7.5.

The daily weather cycle

At night, air near the ground cools faster than that at a height of a few hundred metres, creating a temperature inversion (an increase in temperature with height). This suppresses vertical mixing and results in a stratification of air pollutants. An inversion also slows the wind speed near the ground, and effluents from even tall chimneys are trapped aloft (Munn & Bolin, 1971). During stormy weather or in broken terrain, this cap of polluted air is disrupted and pollutants may diffuse upwards as far as the troposphere. Sometimes hot chimney effluents with a high exit velocity, perhaps containing large amounts of steam, may penetrate upwards through the top of the ground-based inversion, allowing natural atmospheric dispersion and chemical transformation processes to become effective.

After sunrise, warm surface-level convective air bubbles develop upwards until they mix with cooled chimney effluents that have been trapped during the night. These plumes then mix downwards to ground level causing fumigation and provide one of the reasons for the typical morning peak in pollutant concentrations. However, increased pollutant concentrations resulting from turbulent mixing (due to superadiabatic lapse rate) are of short duration since the condition is unstable. The hourly fluctuations in both urban and regional pollutant concentrations result from an imbalance between pollution production rates from, for instance, vehicle exhausts and industrial stacks, and loss rates by vertical mixing through the surface atmospheric layer, transport by wind out of the local air space, chemical reactions within the air space and ground absorption and precipitation scavenging (Munn & Bolin, 1971).

Boundary layers

Boundary layers, or the points of discontinuity between different air-flow patterns, may develop on a micro- or macro-scale. A single leaf, a single building and certainly the leading edges of cities may all develop boundary layers. A boundary layer develops at the leading edge of the new type of surface encountered by the air flow and thickens downwind. On a large scale such effects may be found at the junction of water and land and at the junction of open countryside and forest.

The urban heat island often generates a local wind circulation with warm air rising over the centre of the city and a compensating drift of cooler countryside air moving at lower levels into the urban area.

Problems may occur when a tall structure is erected in the vicinity of an adequately functioning chimney or ventilation shaft. A change in flow patterns and altered ground turbulence may increase down-draught and result in impaired air quality.

Fumigation may occur when the lapse rate is neutral below a chimney, with a stable lapse rate above. Fumigation due to looping develops when there exists a superadiabatic lapse rate, but under this condition pollutant concentrations are of short duration at ground level since there is a high degree of turbulence. Of particular concern for air quality managers is the fact that local wind circulation may result in convergence zones within which trapped pollutants concentrate.

7.2 ACIDIFICATION PATHWAYS

A major environmental concern is the acidification of water bodies as a result of acid precipitation. Among the principal sources of 'acid rain' are the oxides of sulphur and nitrogen. Depending on dispersion patterns, emission of these compounds from coal-fired power plants and vehicle exhausts may pose both local and widespread regional hazards.

Durham (1982) has summarized the most significant pathways for the acidification of sulphur and nitrogen compounds in the atmosphere, as shown in Fig. 7.2. The reactions are categorized either as homogeneous (gas phase) or heterogeneous (liquid phase and particulate) types. A historical review of the scientific understanding of ecosystem acidification has been published by Gorham (1989), and a global perspective provided by Rodhe (1989).

7.2.1 Sulphur dioxide

Sulphur dioxide is a fairly soluble gas and under wet conditions can be removed as sulphurous acid or by reactions in the liquid phase. Dry removal is also believed significant (Donovan, 1983) and the rate determining step is thought to be oxidation to form sulphur trioxide, which reacts rapidly with water to yield sulphuric acid.

Long-term changes in pH of forest soils in southern Sweden have been reported by Falkengren-Grerup (1987) and peat acidification has been linked with acid deposition in Scotland (Skiba *et al.* 1989). The adverse effects of aluminium on developing brook trout (*Salvelinus gontinalis*) in low-calcium water (Hunn *et al.*, 1987) are typical of the ecotoxicological consequences of surface water acidification in poorly buffered lakes. The application of neutralizing material such as limestone to acidic waters has been discussed by Simonin (1988), and the implications of acid rain on agricultural production have been considered for Ontario, Canada (Ludlow & Smit, 1987). Sulphate deposition to surface waters in Norway and the eastern USA has been reviewed by Henriksen and Brakke (1988), and a method for the estimation of historical sulphate concentrations in natural freshwaters has been proposed by Epstein (1988). For the Regional Integrated Lake Watershed Acidification Study data set, the latter author was able to accurately predict sulphate concentrations in 20 out of 23 water bodies by using the following equation:

$$[SO_4^{2-}] = (\text{specific conductance} - 0{\cdot}102\,[HCO_3^-] - 0{\cdot}134[Cl^-] - 0{\cdot}129[NO_3^-] - 0{\cdot}292[H^+] + k)/0{\cdot}138$$

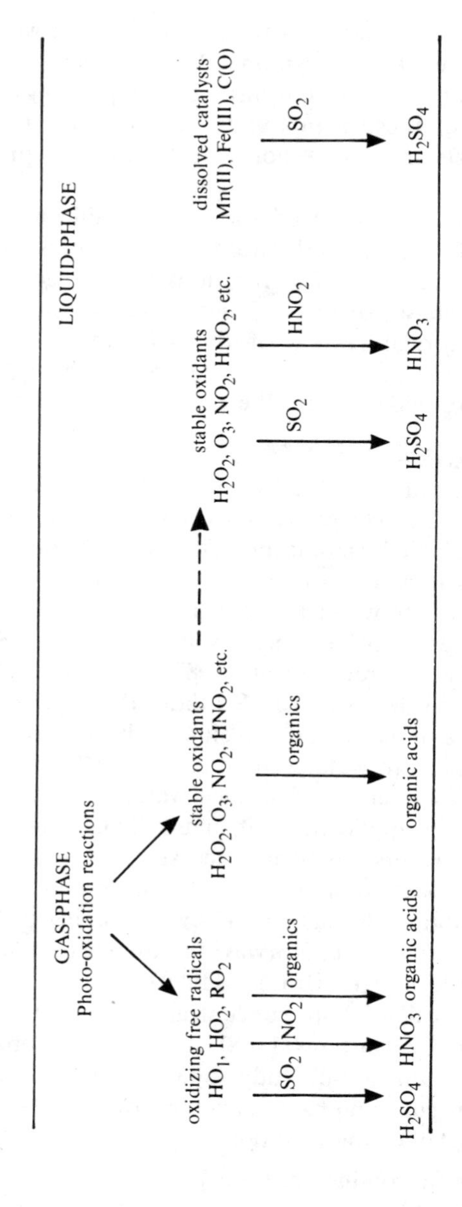

Fig. 7.2. Major atmospheric acidification pathways. Reproduced from Summers (1983) (after Durham (1982)).

where specific conductance was expressed in micromhos per centimetre; all concentrations were in microequivalents per litre and k was a constant reflecting the concentrations of sodium and potassium generally found in the water.

The United Nations Environment Programme (UNEP) Global Environment Monitoring System (GEMS) has, since 1974, been monitoring air quality in about 35 countries. Data on suspended particulate matter and sulphur dioxide (SO_2) are collected from primarily urban locations. Figure 7.3 presents a summary of the annual SO_2 averages recorded between 1980 and 1984 (GEMS/WHO, 1988). An earlier publication reported on trends between 1973 and 1980 (GEMS/WHO, 1984). The most commonly occurring values were between 20 and 60 $\mu g\,m^{-3}$; 30% of cities had higher values, with the highest annual average SO_2 level being 185 $\mu g\,m^{-3}$. Cities in developing countries were represented throughout the entire range of reported average values. The guideline published by WHO (1979) specified a range of 100 to 150 $\mu g\,m^{-3}$ for the 98 percentile of daily average concentrations, which was designed to ensure that even very sensitive members of the population should be protected from short-term adverse effects. A separately derived guideline of 40 to 60 $\mu g\,m^{-3}$ annual mean was intended to avoid any risk of increased respiratory illness in children or of increased prevalence of respiratory symptoms in both children and adults through long-term exposure. Annual mean SO_2 concentrations in Greater London, UK, have been reported at 40 $\mu g\,m^{-3}$ in recent years (Laxen & Thompson, 1987).

7.2.2 Oxides of nitrogen

Oxides of nitrogen (NO_x) from combustion sources such as vehicles and heating installations are a significant component of atmospheric pollution. Reactions of these oxides in the atmosphere give rise to photochemically produced substances such as ozone and peroxyacetyl nitrate (PAN). These, and the pollutant nitrogen dioxide (NO_2), can have a detrimental effect on human health and the environment. In London, UK, NO_2 concentrations have been monitored by diffusion tubes, which are passive devices for measurement over a period of days or weeks, rather than hours (GLC/SSB, 1985/6). The 1987 European Council Directive on Air Quality Standards for Nitrogen Dioxide, which came into force in July 1987, set a limit of 200 $\mu g\,m^{-3}$ as a 98 percentile and a guide value of 135 $\mu g\,m^{-3}$ as a 98 percentile

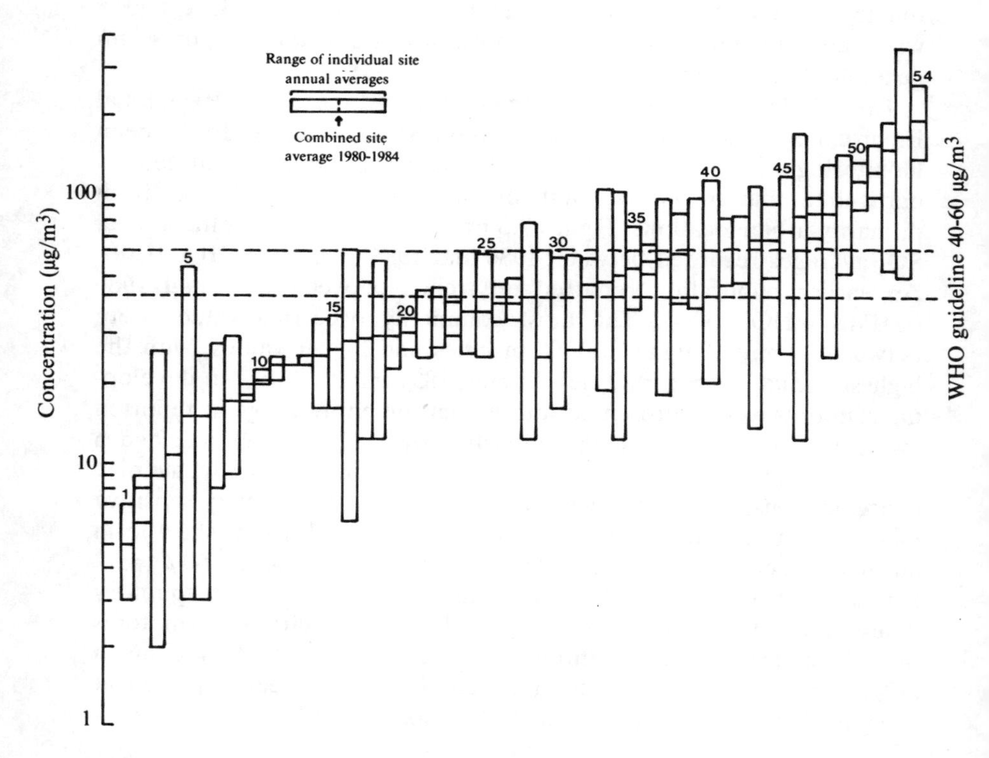

Fig. 7.3. Summary of the annual SO_2 averages in GEMS/Air cities, 1980–84. 1, Craiova; 2, Melbourne; 3, Auckland; 4, Cali; 5, Tel Aviv; 6, Bucharest; 7, Vancouver; 8, Toronto; 9, Bangkok; 10, Chicago; 11, Houston; 12, Kuala Lumpur; 13, Munich; 14, Helsinki; 15, Lisbon; 16, Sydney; 17, Christchurch; 18, Bombay; 19, Copenhagen; 20, Amsterdam; 21, Hamilton; 22, Osaka; 23, Caracas; 24, Tokyo; 25, Wroclaw; 26, Athens; 27, Warsaw; 28, New Delhi; 29, Montreal; 30, Medellin; 31, St Louis; 32, Dublin; 33, Hong Kong; 34, Shanghai; 35, New York; 36, London; 37, Calcutta; 38, Brussels; 39, Santiago; 40, Zagreb; 41, Frankfurt; 42, Glasgow; 43, Guangzhou; 44, Manila; 45, Madrid; 46, Beijing; 47, Paris; 48, Xian; 49, Sao Paulo; 50, Rio de Janeiro; 51, Seoul; 52, Tehran; 53, Shenyang; 54, Milan. Reproduced from GEMS/WHO (1988).

with 50 $\mu g\,m^{-3}$ for the 50 percentile. The Directive did not apply to exposure at work or inside buildings.

Nitric oxide (NO) and NO_2 must be considered together as they are rapidly interconverted in the atmosphere (Donovan, 1983). During daylight, NO_2 is photolysed to yield an oxygen atom and NO. The oxygen atom is then combined with an oxygen molecule (O_2) to form ozone (O_3), which can return to react with NO and form NO_2 again. Thus, under conditions of bright sunlight the main species present in the atmosphere will be NO, while under conditions of low light intensity the predominant species will be NO_2.

Nitrous oxide (N_2O) is relatively inert in the troposphere, although increasing use of nitrate fertilizers may lead to increased rates of N_2O release, to the possible detriment of the ozone cover. The main removal process for NO_x species in the troposphere is reaction with OH radicals to form nitric acid (HNO_3), which is subsequently removed by rain. The classical photochemical smog cycle requires:

(i) strong sunlight;
(ii) stable meteorological conditions;
(iii) the presence of NO_x;
(iv) the presence of unsaturated hydrocarbons.

The cycle starts with hydrocarbon and NO emissions from car exhausts. The hydrocarbons are then attacked by hydroxyl radicals to yield eventually RO_2 radicals. The RO_2 radicals can then oxidize the NO to NO_2, which in turn is photolysed to yield an oxygen atom and ultimately, on combination with O_2, an ozone molecule. Once ozone is formed a rapid series of complex reactions follow, which lead to the formation of aerosols and the so-called smog.

The analysis of the characteristics of complex chemical reaction mechanisms with reference to photochemical smog chemistry has been considered by Leone and Seinefeld (1984), and Atkinson *et al.* (1987) have reported on the kinetics of the reactions of NO_3^- radicals with a series of aromatic compounds. The interaction of soluble oxidants and organic toxins with SO_2 and NO_x has been reviewed by Gaffney *et al.* (1987).

Measurements made since the 1850s have shown that precipitation contains hundreds of organic compounds (Mazurek & Simoneit, 1986), thus, considering only hydrogen ions, sulphate, nitrate and trace inorganic elements in contaminated precipitation is a gross oversimplification of the reality in ecosystems. Cloudwater and rainwater samples were collected from 10 nonurban sites in North America over two years by Weathers *et al.* (1988). It was found that on average, cloudwater collected from sites in the eastern USA was more acidic and had higher concentrations of NO_3^- and SO_4^{2-} than at sites in the western USA and Puerto Rico. Concentrations of H^+, NO_3^- and SO_4^{2-} were significantly higher in cloudwater than in rainwater at most sites, although on a paired-event basis, enrichment factors for cloud versus rainwater varied considerably. In contrast to distributions of inorganic ions, the concentrations of formate and acetate in cloud-water and rainwater were similar at sites in both the eastern and western USA, suggesting natural rather than anthropogenic sources.

7.2.3 Chlorofluorocarbons and ozone depletion

The chlorofluorocarbons (CFCs) were used in increasing quantities throughout the 1960s and early 1970s as aerosol spray propellants, refrigerants, solvents and plastic foaming agents, and have been accumulating steadily in the atmosphere, mainly as $CFCl_3$ and CF_2Cl_2 (Donovan, 1983). Although more or less inert in the troposphere, when transported to the stratosphere these compounds are photolysed to react with electronically excited oxygen atoms to yield chlorine atoms and chlorine oxides (collectively termed ClO_x). Such degradation products form the basis of a further catalytic cycle, analogous to the NO_x cycle, which leads to the net removal of ozone and oxygen atoms.

In the mid 1970s, Molina and Rowland (1974) hypothesized that CFC molecules might be implicated in potential damage to stratospheric ozone. The main expectations outlines were that CFCs would:

1. Survive for many decades in the Earth's atmosphere.
2. Decompose after reaching the mid-stratosphere at altitudes where they are exposed to short-wavelength radiation (190–230 mm), which does not penetrate to lower altitudes because of absorption by ozone (O_3) or by molecular oxygen (O_2).
3. Release atomic chlorine (Cl) that would then initiate long, ozone-depleting ClO_x chains involving Cl and ClO as alternating reactants:

$$Cl^{\cdot} + O_3 \rightarrow ClO + O_2$$

$$ClO + O \rightarrow Cl + O_2$$

The ClO_x chains have been known since 1974 to be most effective in the depletion of O_3 through reactions at altitudes between 25 and 45 km, in which future ozone losses with continued emission of CFCs have been predicted in the range of 35 to 60% (Rowland, 1988). It should be recognized that ozone is actually only a trace constituent of the stratosphere; at a maximum concentration, ozone only makes up a few parts per million of the air molecules. If the diffuse ozone layer were concentrated into a thin shell of pure ozone gas surrounding the Earth at atmospheric pressure, it would measure only about 3 mm in thickness (Pimentel, 1986). Furthermore, ozone destruction mechanisms are based on chain reactions in which the pollutant molecules may destroy many thousands of ozone molecules before being transported to the lower atmosphere, chemically transformed and removed by rain. Currently, it is recognized that the rates of some 150 chemical reactions must be considered in order to approach a quantitative model for simulating the stratosphere and predicting changes resulting from the introduction of various pollutants. Figure 7.4 illustrates both how NO and NO_2 reactions together furnish a true catalytic cycle in which NO and NO_2 are the catalysts, and the cycle for CF_2Cl_2.

The main reasons for concern focus on the twin roles of ozone as a stratospheric shield against the penetration to the Earth's surface of biologically damaging ultraviolet radiation in the 280–320 nm wavelength band (designated as UV-B), and the conversion of this energy absorbed by ozone into a stratospheric heat source. A decrease in total ozone in the atmosphere permits increased penetration of UV-B in the ratio of -1% in O_3 equalling a 2% increase in UV-B (Rowland, 1988). The major direct effects of increased UV-B on humans are increased incidences of skin cancer and eye cataracts, and perhaps some suppression of the immune system (NAS, 1984).

The Antarctic ozone hole

Between 1977 and 1984 the springtime amounts of ozone in the atmosphere over Halley Bay, Antarctica, decreased by more than 40% (Farman *et al.*, 1985). Other groups soon confirmed the report and showed that the region of ozone depletion was actually wider than the continent and extended roughly from 12 to 24 km in altitude, spanning much of the lower stratosphere. There was, in essence, an ozone 'hole' in the polar atmosphere (Stolarski, 1988).

A search for the possible causes of this springtime Antarctic ozone loss began, with contending theories being sorted into classes of

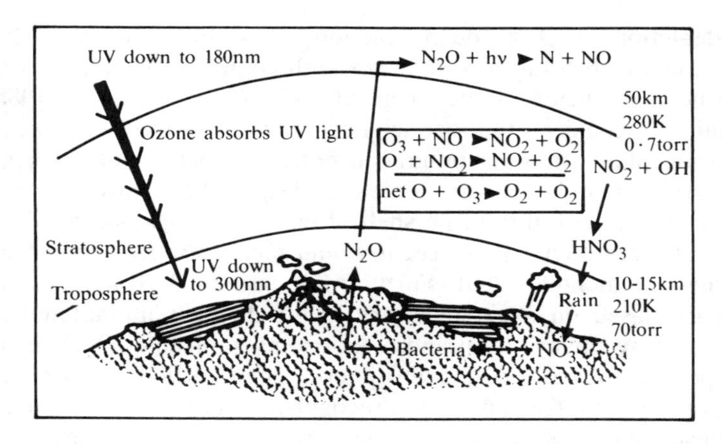

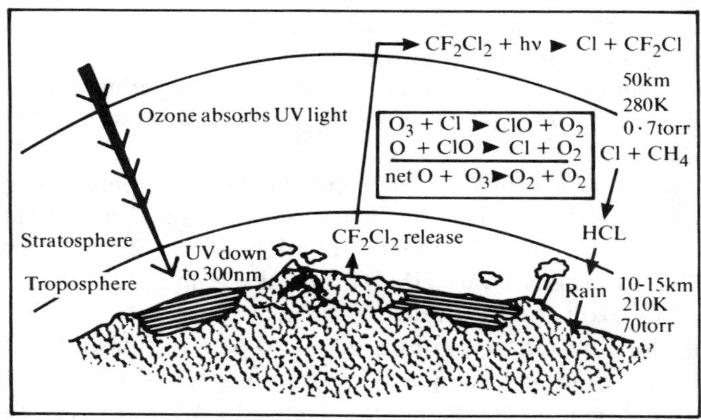

Fig. 7.4. Some stratospheric ozone destruction mechanisms. Reproduced from Pimentel (1986).

chemical versus dynamic, and natural versus anthropogenic (Rowland, 1989). Scientific expeditions in 1986 and 1987 (NASA, 1988) demonstrated unequivocally that the ozone hole is the consequence of chlorine chemistry in conjunction with the special conditions of Antarctic meteorology.

As a result of these findings, a United Nations Environment Programme convention to protect the ozone layer was signed in Vienna in 1985, and a protocol outlining proposed protective actions followed in Montreal in September 1987. The protocol specified a 20%

reduction from 1986 emissions of fully halogenated CFCs by 1994 and a further 30% reduction by 1999. Unfortunately, a large loophole exists in the protocol allowing developing countries to go into production of CFCs. Even if the protocol were fully effective, the reductions are likely to have only a minor effect before the end of the century (Rowland, 1989).

7.3 ATMOSPHERIC SINK MECHANISMS

Large or coarse particles with diameters greater than 2·0 to 2·5 μm are produced mainly by mechanical means such as Aeolian weathering of soils, sea spray, volcanic activity, and release from plants as, e.g. pollen and spores. The smallest particles, of diameter less than 0·08 μm are known as Aitken nuclei which arise from gas-to-particle conversion. The smallest diameter range contains most of the total number of particles, but little mass. Gas-to-particle conversion and coagulation of Aitken nuclei produces particles of intermediate size, the accumulation mode (Bidleman, 1988).

Most of the surface area and about half the mass of urban air particulate matter (Whitby, 1978) is represented by the accumulation mode. These particles are too small to undergo rapid gravitational settling and are slowly removed by rain and dry deposition. The major constituents of urban air particulate matter are organic and elemental carbon, sulphate, nitrate, ammonium, silicates, alkali and alkaline earth metals, aluminium, iron and lead. Total carbon comprises between 10 and 20% of the total suspended particulate load (Shah *et al.*, 1986). Reviews of the carbonaceous fraction include those by Daisey (1980), Simoneit and Mazurek (1981), Duce *et al.* (1983) and Lioy and Daisey (1986).

7.3.1 Removal processes

Sink mechanisms provide the means of explaining and quantifying rates of change of pollutant concentration in the troposphere. Depending on the nature of the substance, sink mechanisms range from stratospheric hydroxyl radical reactions to stomatal intake on plant leaves.

Important processes for the removal of *particles* from the troposphere were itemized by Whelpdale and Munn (1976) as follows:

(a) Wet removal by precipitation.
(b) Dry removal by sedimentation.
(c) Dry removal by impaction on vegetation.

Important processes for the removal of *gases* from the troposphere are:

(a) Wet removal by precipitation.
(b) Absorption or reaction at the Earth's surface.
(c) Conversion into other gases or particulates by chemical reaction within the atmosphere.
(d) Transport into the stratosphere.

Wet removal by precipitation, or precipitation scavenging, is one of the most effective atmospheric cleansing mechanisms for both particles and gases. Where chemicals are incorporated into precipitation elements within clouds the process is referred to as 'rainout' or 'snowout', and where incorporation occurs below clouds the processes are termed 'washout'. The rainfall 'washout ratio' (W) can be calculated relatively simply as the reciprocal of the Henry's Law constant (H) and also measured experimentally as

$$W = \frac{\text{Mass of substance . litre}^{-1}\text{ rain}}{\text{Mass of substance . litre}^{-1}\text{ air}}$$

The processes of dry and wet deposition have been described qualitatively by Whelpdale (1983) and, for the major acidic species, summarized by Whelpdale (1982) as shown in Fig. 7.5. Detailed technical reviews of the processes have been published by Garland (1978), Hales (1978), Liss and Slinn (1983), Pruppacher *et al.* (1983), Buat-Menard (1986) and Bidleman (1988).

Particle removal by gravitational sedimentation is an effective process only for particles of radius larger than approximately 10 μm; the fall velocity of smaller articles becomes insignificant in comparison with atmospheric vertical motions (Whelpdale & Munn, 1976). The scavenging efficiency of vegetation is dependent on factors such as particle size and the morphology, area and wetness of collecting surfaces. Increasingly cited as an important factor, both in terms of environmental fate and efficiency of application, is rain washoff of pesticides from foliage and soil (Cohen & Steinmetz, 1986).

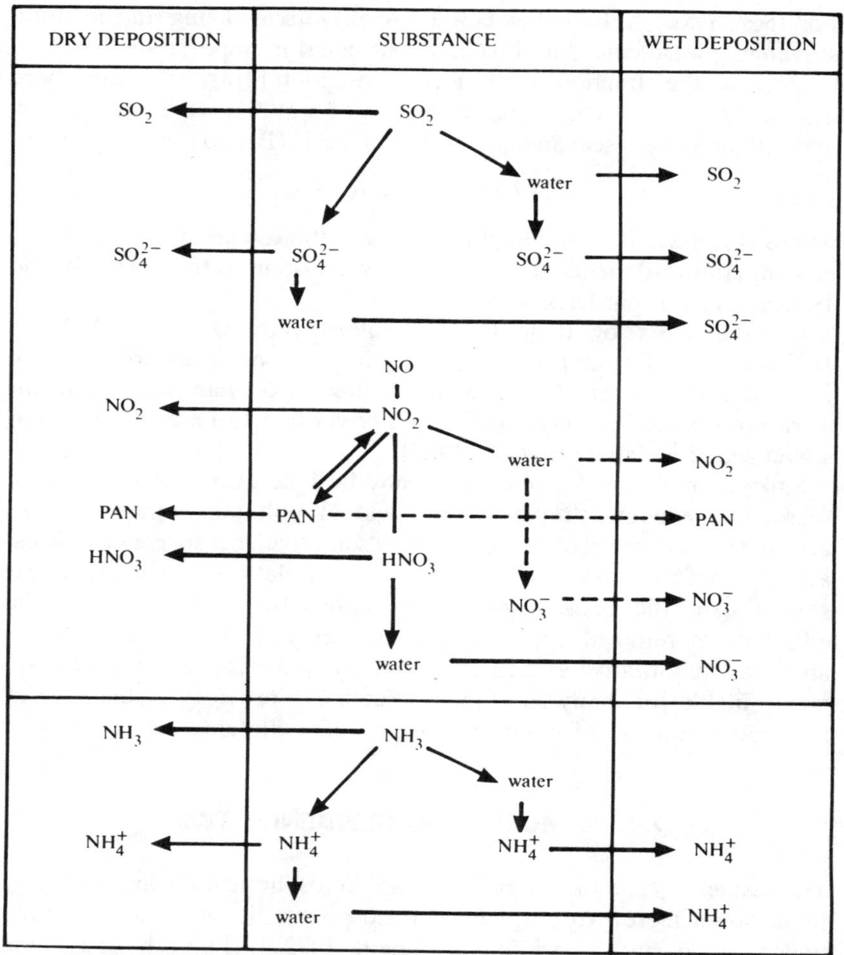

Fig. 7.5. Schematic diagram of possible deposition pathways for the major acidic species. (The middle column depicts processes or steps along the pathway prior to deposition.) ⇄ Important pathways; ⇠⇢ Uncertain pathways. Reproduced from Whelpdale (1982).

Scavenging of airborne PAHs by rain was studied by van Noort & Wondergem (1985) who concluded that the main process responsible for the presence of phenanthrene in rain is below-cloud gas-phase scavenging, while in-cloud scavenging was believed to be the main process for other PAHs except fluoranthene and benz[*a*]anthracene.

For these two PAHs, below-cloud gas-phase scavenging and in-cloud scavenging were considered to be about equal in importance.

If ϕ is the fraction of a given compound in the atmosphere associated with aerosol particles (Junge, 1977), then to a first approximation gas scavenging will dominate if (Pankow *et al.*, 1984):

$$RT/H(T) \gg W_P\phi/(1-\phi)$$

where R is $8{\cdot}2 \times 10^{-5}$ m^3 atm/(mol K), T is the temperature (K), $H(T)$ is temperature-dependent Henry's Law constant (atm mol m^{-3}) and W_P are washout particles.

On the basis of their low H values (approximately 10^{-7} to 10^{-6} atm m^3 mol^{-1}) and their relatively large vapour pressures (10^{-3} to 10^{-1} torr) and therefore low ϕ values, low molecular weight phenols were both predicted and experimentally verified to be exclusively gas scavenged (Leuenberger *et al.*, 1985).

Sink estimates on a global scale may be inaccurate for a variety of reasons. Frequently 'representative global pollutant concentrations' are merely averages of a few unrepresentative measurements. Most available information refers to land surface data as there are fewer studies over the oceans and in the upper troposphere. Often the efficiency of removal by, say, forest scavenging, is not easily calculable, and quantitative estimates of washout and rainout coefficients are not available for many substances. Table 7.1 provides a summary of information on sinks for various gaseous air pollutants.

7.4 ATMOSPHERIC RESIDENCE TIME

The residence time of a chemical in an atmospheric compartment (e.g. total atmosphere, troposphere, stratosphere) is well defined only under steady-state conditions (Lyman, 1982*b*). In such cases, the residence time (τ) may be simply defined as the ratio of the total mass in the compartment (Q) to the total emission rate (E) or removal rate (R) where

$$\tau = Q/E = Q/R$$

In this equation, E is considered to be the sum of all land, fresh water and ocean emissions to the atmosphere, plus any other inputs, including *in-situ* atmospheric generation. Similarly, R is the sum of losses from all compartments and losses by *in-situ* degradation.

Table 7.1
Sinks for selected gaseous air pollutants

Substance	*Sinks*
Sulphur dioxide	Precipitation scavenging: washout, rainout Oxidation in gas and liquid phase to sulphate Soil: microbial degradation, physical and chemical reaction, absorption Vegetation: sorption on surfaces, stomatal intake Oceans, lakes: absorption
Hydrogen sulphide	Oxidation to sulphur dioxide
Ozone	Chemical reaction on vegetation, soil, snow and ocean surfaces
Nitrous oxide	Soil: microbial destruction Stratosphere: photodissociation Ocean: absorption
Nitric oxide/ nitrogen dioxide	Soil: chemical reaction Vegetation: sorption, stomatal uptake Chemical reaction in gas and liquid phase
Ammonia	Chemical reaction to ammonium in gas and liquid phase Precipitation scavenging: washout, rainout Surface uptake: physical and chemical reaction, absorption
Carbon monoxide	Stratosphere: reaction with hydroxyl radical Soil: microbiological activity
Carbon dioxide	Vegetation: photosynthesis, absorption Oceans: absorption
Methane	Soil: microbiological activity Vegetation: chemical reaction, bacterial action Troposphere and stratosphere: chemical reaction
Hydrocarbons	Chemical reaction to particulates Soil: microbiological activity Vegetation: absorption, stomatal intake

Data from Whelpdale and Munn (1976).

Atmospheric residence time is also termed 'turnover time' or 'average transit time' and has been more formally defined by Bolin and Rodhe (1973) and Slinn (1978).

When the removal rate (R) for a chemical is due solely to some first-order loss process, then the half-life ($t_{1/2}$) and τ are related as follows:

$$t_{1/2} = 0{\cdot}693\tau$$

τ should not be confused either with the 'average age' of a chemical molecule in the atmospheric compartment or with the half-life of the chemical.

Atmospheric residence time is not an intrinsic chemical property, nor is it well defined for a given chemical in a specified compartment. It merely provides a rough, temporally and spatially averaged measure of the input fluxes and removal processes acting on a chemical in an artificially and arbitrarily defined atmospheric compartment. Factors influencing residence time may include latitude, input flux and atmospheric phenomena. The latter are often associated with characteristic time scales, e.g. seasonal cycles of precipitation and temperature. Some other time scales are listed in Table 7.2.

Atmospheric residence time cannot be measured directly. Instead it must be calculated or inferred on the basis of simplified atmospheric models. Lyman (1982*b*) suggested that only the imagination of the modeller, the data available and the computational facilities to hand

Table 7.2
Time scales for atmospheric phenomena

Process	*Typical time scale*	*Reference*
Precipitation or nucleation scavenging	1 week	Junge (1977), Slinn (1978)
Vertical mixing time of troposphere	1 week	Slinn (1978)
Horizontal mixing time of troposphere	1 year	Slinn (1978)
Mixing between northern and southern hemispheres	1 year	Singh (1977*b*), Singh and Salas (1979)
Movement from troposphere to lower stratosphere	4 years[a]	Junge (1977)
Movement from lower stratosphere to troposphere	1 year[a]	Junge (1977)

[a] Time required for exchange of air between the specified compartments. Movement from the troposphere to the lower stratosphere takes longer than the reverse process because the troposphere contains about four times as much air as the stratosphere.

limit the number of ways by which residence time can be estimated. However, the latter author cited five methods, which can be solved without a computer and with limited data, namely:

1. Steady-state model (Bolin & Rodhe 1973; Slinn, 1978)
2. Non steady-state, unicompartmental model (Singh 1977*a*; Singh *et al.*, 1978)
3. Non steady-state, bicompartmental model (Singh, 1977*b*; Singh *et al.*, 1978)
4. Using chemical reactivity data (Brown *et al.*, 1978; Slinn, 1978)
5. Correlation with mean standard deviation (Junge, 1974).

Applicable to both organic and inorganic chemicals, these methods are intended only for calculating tropospheric residence times. Data required for estimations are listed in Table 7.3, but precise details of the methods should be sought in Lyman (1982*b*) or in the sources cited. The residence times of 'heavy' metals in the environment were discussed by Bowen (1975).

7.5 ATMOSPHERIC DISPERSION MODELLING

7.5.1 Simple models

Roll back models

Roll back models are the simplest of air quality simulation models. They are based on the assumption that the local concentration of a pollutant above its background level is directly proportional to the emission from a given source. Although these models have the advantage of being very easy to apply, they are *not* applicable to *reactive* pollutants.

Box models

Box models assume that pollutants are uniformly mixed throughout a fixed volume (box) of air. The box is taken to extend vertically to the inversion base. Concentrations are then presumed to be proportional to the rates of source emission and inversely proportional to the average residence time and the inversion height.

Table 7.3
Data required for estimation of atmospheric residence time

Method		*Required data*
Steady-state model	(1)	Average concentration of chemical in troposphere (C); this is used to estimate total mass of chemical in troposphere (Q)
	(2)	Rate of emission of chemical into troposphere (E) *or* Rate of removal of chemical from troposphere (R)
Non steady-state, one-compartment model	(1)	Average concentration of chemical in troposphere (C); this is used to estimate total mass of chemical in troposphere (Q).
	(2)	Year-by-year emissions inventory for chemical; this is used to obtain cumulative emissions (A) and the parameter (b) in the exponential expression for the rate of emission in recent years
Non steady-state, two-compartment model	(1)	Average concentrations of chemical in both northern and southern hemispheres; these are used to estimate the total mass of chemical in the northern (Q_N) and southern (Q_S) hemispheres
	(2)	Year-by-year emissions inventory for chemical; this is used to obtain cumulative emissions (A) and the parameter (b) in the exponential expression for the rate of emission in recent years
	(3)	Interhemispheric exchange rate (τ_e); this may be taken as $\approx$1·2 years
Use of chemical reactivity data	(1)	Rate constants for reaction of chemical with hydroxyl radical (k_{OH}), ozone (k_{O_3}) and other reactants, if any
	(2)	Concentration of hydroxyl radical, [$OH^{\cdot}$], ozone [O_3], and any other reactant being considered
Correlation with mean standard deviation (Junge's correlation)	(1)	Average concentration of chemical in troposphere (C).
	(2)	Standard deviation (σ) associated with average concentration

Data from Lyman (1982*b*).

Gaussian plume models

Although a wide variety of dispersion models have been developed, there is no doubt that US Environmental Protection Agency (EPA) models have attained a certain 'regulatory status' (Budiansky, 1980). These models are all based on a Gaussian (binormal or bell-shaped) plume spread. The Gaussian distribution of the concentration of a substance is the general solution to the diffusion equation when the simplifying assumption of 'Fickian' diffusion is made. The 'K theory' equation relates the concentration of a substance as a function of position and time, to the action of average wind flows and turbulence. The K theory equation in one dimension is

$$\frac{\partial C}{\partial t} = -\frac{\partial}{\partial x}(uC) + \frac{\partial}{\partial x}\left(K\frac{\partial c}{\partial x}\right)$$

where C is the concentration of the substance, u is the wind speed in the x direction, K is a coefficient known as the 'turbulent diffusivity' or 'eddy diffusivity', x is the distance from the source and t equals time.

The first term on the right-hand side of the equation represents transport due to average winds; the second term represents diffusion due to turbulence.

Four assumptions must be made to obtain the Gaussian solution:

1. The solution is time-invariant.
2. Wind speed is not a function of position.
3. Diffusivities are not functions of position (the key assumption in 'Fickian' diffusion).
4. Diffusion in the x direction is insignificant compared with mean flow in the x direction.

Given such assumptions, the equation has an analytical solution, which expresses the concentration as a Gaussian, or normal, distribution in the y and z directions. The Gaussian solution:

$$C(x, y, z) = \frac{Q}{\pi u \sigma_y \sigma_z} \mathrm{e} - \left[\frac{y^2}{2\sigma_y^2} + \frac{z^2}{2\sigma_x^2}\right]$$

where C is the concentration, Q is the emission rate at the source, u is the wind speed in the x direction, σ_y is a function of the turbulent diffusivity in the y direction and the distance x, σ_z is a function of the turbulent diffusivity in the z direction and the distance x, x is the horizontal distance from the source perpendicular to u, and z is the vertical distance from the source.

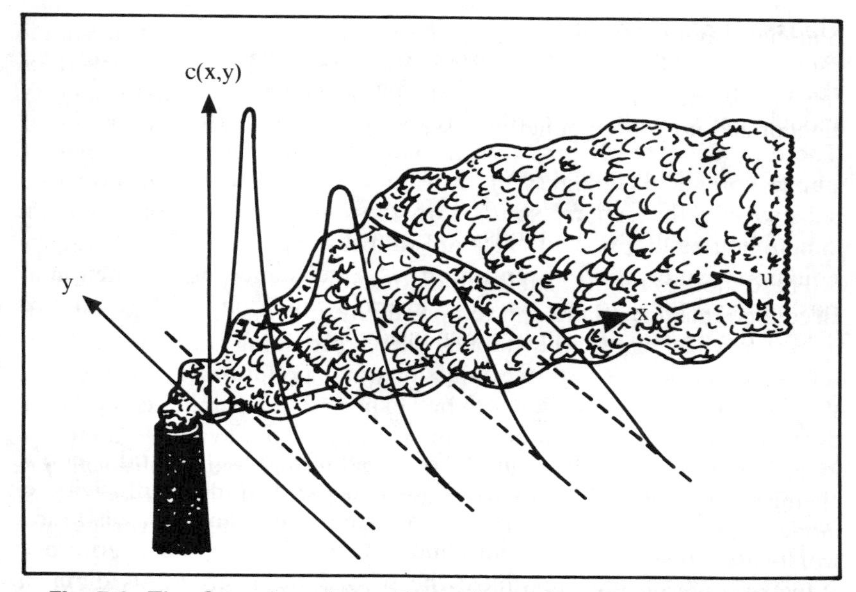

Fig. 7.6. The Gaussian plume. Reproduced from Budiansky (1980).

Figure 7.6 illustrates the characteristic horizontal spread of a Gaussian plume as it travels downwind from the source. Concentration is shown on the vertical axis as a function of x and y for a fixed value of z.

Errors in predicted concentration may range from 30% to one order of magnitude, depending on the compound, the averaging time, the spatial scale, the terrain and the precise choice of model. None the less, Gaussian models are commendable for their relative simplicity, requiring little computational time and few input data. A major drawback to Gaussian models is that they are all steady-state, with factors such as wind speed, temperature, emission rates and mixing height taken as constants. A second source of error is the empirical formulation used to specify the horizontal and vertical dispersion parameters, which determine how the plume spreads as a function of distance from the source.

7.5.2 More complex models

Photochemistry is a major factor in determining the concentrations of ozone and nitrogen oxides in the lower atmosphere, and since this is a time-dependent phenomenon it cannot be incorporated into the simple steady-state Gaussian model. More 'complex' models, all of them relying on numerical solutions of the K-theory equation, may be categorized into three general classes:

A. Eulerian formulations.
B. Lagrangian formulations.
C. Particle-in-cell methods.

These models require substantial computer capacity and a relatively large amount of spatially and temporally resolved source emission and meteorological data. Inevitably, while providing insight into the relationships between source emissions and pollution at the receptor site of interest, they cannot be expected to simulate pollutant behaviour with great accuracy. This is because the available meteorological data are usually inadequate and because many physical and chemical transformation processes are still not well understood. Thus, model runs are generally limited to the prediction of average and worst-case situations.

In the *Eulerian* formulation, a fixed coordinate system, or grid, is laid out over the entire region of interest. The concentration of a pollutant in each square of the grid is then calculated by explicitly solving the equation, using numerical methods and the help of a computer, over a series of small time intervals. In squares that contain sources, a source term is added to the right-hand side of the K-theory equation. This approach is most useful for situations in which there are multiple sources or for which predicted concentrations are needed for the entire region. The model of Prahm and Christensen (1977) was used fairly successfully to explain daily averages of sulphate measured in Europe.

The *Lagrangian* formulation, on the other hand, considers a coordinate system moving with the local mean winds. This approach is particularly useful for modelling transport over a long distance or when the effect on a particular receptor site only is of concern—in other words, when there is no need to calculate concentrations over an extensive array of fixed locations. While this saves on computer memory and time, it leads to results that can be difficult to interpret, since the coordinate grid is distorted as it twists and turns with the local winds. A subset of Lagrangian models, *trajectory* models, apply the Lagrangian formulation to a single moving cell, thus avoiding the grid distortion. In this case, however, turbulent diffusion must be calculated from external data, and is in practice usually neglected altogether.

The Lagrangian model developed by Eliassen and Saltbones (1975) to help explain the causes of acid rain in Scandinavia is typical of early versions of such models. It is based on the evolution of the mass of material enclosed with a control volume, and treats sulphur only.

The *particle-in-cell* method is a hybrid approach. Here, the source emissions are divided into individual Lagrangian cells, each of which is tracked over a fixed coordinate system. The concentration in each fixed grid square is then calculated simply by counting up the number of these cells present in each square (Budiansky, 1980).

The principal features of some atmospheric dispersion models are outlined in Fig. 7.7. Keddie (1983) provided a useful guide to atmospheric dispersal of pollutants and the modelling of air pollution, and Turner (1979) and Fisher (1983) comprehensively reviewed the subject of atmospheric dispersion modelling and long-range transport of air pollutants. Eschenroeder (1975) summarized some model intercomparisons, including those between photochemical models, and Doury (1982) discussed the limitations to simple mathematical formulae used in atmospheric dispersion studies. Meteorological aspects of global air pollution were reviewed by Munn and Bolin (1971). Tables 7.4 and 7.5 provide a guide to some selected air pollution models. Venkatram and Karamchandani (1986) have reviewed some of the acid deposition models mentioned in Table 7.4, and Ellis (1988) has considered acid rain control strategies in the context of some additional long-range transport models.

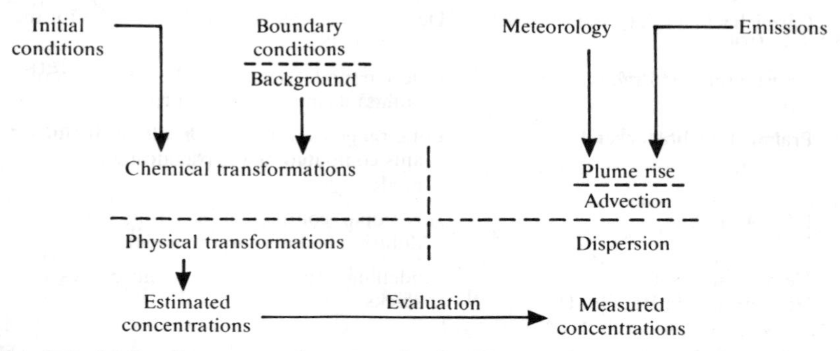

Fig. 7.7. Main elements of atmospheric dispersion models. Reproduced from Turner (1979).

Table 7.4
Selected air pollution models

Reference	*Model details*
Rodhe (1972) McMahon *et al.* (1976)	Simple 'box' model to calculate sulphur deposition over Europe
Sheih (1977) Bolin and Persson (1975)	Used simple Gaussian distribution to describe the horizontal spread of airborne material over a long period of time
ApSimon and Goddard (1976) ApSimon *et al.* (1980)	An advanced model (MESOS) for investigating the probability distribution of concentration from a nuclear accident release, based on following the trajectories of a series of puffs
Maul (1982*a,b*)	Used a similar type of model to investigate sulphur transport during precipitation episodes over Norway
Bhumralker *et al.* (1981)	A time-dependent model with varying boundary layer depths to investigate the importance of stack heights. The model of Johnson *et al.* (1978) has been extended to include variable mixing height and to allow for shear in the transport wind
Eliassen (1976, 1978)	A trajectory model used in an Organisation for Economic Co-operation and Development (OECD) project
EMEP (1980)	Incorporates improvements to the OECD model above
McNaughton (1980) McNaughton and Scott (1980)	Have developed several advanced trajectory models in the USA, which compare results for sulphur compound monitoring networks
Carmichael and Peters (1979, 1980)	Developed a regional transport model for the eastern USA
Christensen and Prahm (1976)	Eulerian model describing the dispersion of pollutants in gases and fluids
Prahm and Christensen (1976)	Long-range transmission of sulphur pollutants computed by the pseudospectral model
NOAA (1975)	A regional–continental scale transport, diffusion and deposition model
Venkatram (1980, 1981) Venkatram and Vet (1981)	Modelling dispersion from point sources/tall stacks

Table 7.4—contd.

Reference	*Model details*
Dennis (1980)	The development of the smeared concentration approximation (SCA) method. Allows air-pollution dispersion to be included in multidisciplinary studies for long-term policy analysis without the direct use of large, complex air-pollution models. Focuses on urban-area analysis.
Heffter (1980)	Air Resources Laboratories Atmospheric Transport and Dispersion Model (ARL-ATAD)
ADOM/TADAP (1984, 1985)	The Acid Deposition and Oxidant Model (Eulerian) was developed at Environmental Research and Technology under the sponsorship of the government of Ontario, Environment Canada, the FRG government, and the Electric Power Research Institute
NCAR (1985, 1986)	The (Eulerian) Regional Acid Deposition Model (RADM) was developed at the National Center for Atmospheric Research with EPA sponsorship
Carmichael *et al.* (1986)	The Sulfur Transport Eulerian Model (STEM) was developed at the University of Kentucky and Iowa with funding from the US DoE and NASA
Lurmann *et al.* (1985)	PLMSTAR Air Quality Simulation Model

A general integrated model of Transport of Heavy Metals (TOHM), was developed in order to predict the terrestrial fate of zinc, cadmium, chromium, lead and mercury emitted by the operation of a coal-fired electric generating facility (Wagnet *et al.*, 1978; Grenney *et al.*, 1979; Wagnet *et al.*, 1979). TOHM consists of interfaced submodels describing atmospheric dispersion, precipitation, soil chemistry and soil erosion (Fig. 7.8). The models were based on data from a semiarid region of the southwestern USA and, except for climatic and topographic constraints, are not site-specific.

The atmospheric dispersion submodel SPEDTEC requires input data on the effective stack height attained by the effluent plume, the vertical profile of the turbulent intensity of the atmosphere, the presence and location of atmospheric stable layers, the wind field and the particle size distribution of the aerosols. Submodel RAIN in-

Table 7.5
User's Network for Applied Modelling of Air Pollution (UNAMAP)[a]

Model	*Model details*	*Reference*
APRAC	Stanford Research Institute's urban carbon monoxide model. Computes hourly averages for any urban location. Requires an extensive traffic inventory for the city of interest	NTIS (*a,b,c,d*) Dabberdt *et al.* (1973) Johnson *et al.* (1973)
CDM	The Climatological Dispersion Model determines long-term (seasonal or annual) quasistable pollutant concentrations at any ground level receptor, using average emission rates from point and area sources and a joint frequency distribution of wind direction, wind speed and stability for the same period	Busse and Zimmerman (1973)
HIWAY	Computes the hourly concentrations of nonreactive pollutants downwind of roadways. It is applicable for uniform wind conditions and level terrain. Although best suited for at-grade highways, it can also be applied to depressed highways (cut sections)	Zimmerman and Thompson (1975)
	Point source models The three following point source models use Briggs plume rise methods and Pasquill–Gifford dispersion methods as given in EPA's AP-26, *Workbook of Atmospheric Dispersion Estimates,* to estimate hourly concentrations for stable pollutants	Turner and Busse (1973)
PTMAX	Performs an analysis of the maximum short-term concentrations from a single point source as a function of stability and wind speed. The final plume height is used for each computation	
PTDIS	Estimates short-term concentrations directly downwind of a point source at distances specified by the user. The effect of limiting vertical dispersion by a mixing height can be included, and gradual plume rise to the point of final rise is also considered. An option allows the calculation of isopleth half-widths for specific concentrations at each downwind distance	
PTMTP	Estimates the concentration from a number of point sources for arbitrarily located receptor points at or above ground level. Plume rise is determined for each source. Downwind and crosswind distances are determined for each source–receptor pair. Concentrations at a receptor from various sources are assumed to be additive. Hourly meteorological data are used with both hourly concentrations and averages over any averaging time from 1 to 24 h obtainable	

Table 7.5—contd.

Model	*Model details*	*Reference*
CDMQC	This algorithm is the Climatological Dispersion Model (CDM) altered to provide implementation of calibration, of individual point and area source contribution lists, and of averaging time transformations. The basic algorithms to calculate pollutant concentrations used in the CDM have not been modified, and results obtained using CDM may be reproduced using the CDMQC	Brubaker *et al.* (1977)
CRSTER	This algorithm estimates ground-level concentrations resulting from up to 19 colocated elevated stack emissions for an entire year and prints out the highest and second-highest 1-, 3- and 24-h concentrations as well as the annual mean concentrations at a set of 180 receptors (five distances by 36 azimuths). The algorithm is based on a modified form of the steady-state Gaussian plume equation, which uses empirical dispersion coefficients and includes adjustments for plume rise and limited mixing. Terrain adjustments are made as long as the surrounding terrain is physically lower than the lowest stack height input. Pollutant concentrations for each averaging time are computed for discrete, nonoverlapping time periods (no running averages are computed) using measured hourly values of wind speed and direction and estimated hourly values of atmospheric stability and mixing height	NTIS (*e*)
PAL	Point, Area, Line source algorithm. This short-term Gaussian steady-state algorithm estimates concentrations of stable pollutants from point, area and line sources. Computations from area sources include effects of the edge of the source. Line source computations can include effects from a variable emission rate along the source. The algorithm is not intended for application to entire urban areas but for smaller scale analysis of such sources as shopping centres, airports and single plants. Hourly concentrations are estimated and average concentrations from 1 to 24 h can be obtained	Peterson (1978)
Valley	This algorithm is a steady-state, univariate Gaussian plume dispersion algorithm designed for estimating either 24 h or annual concentrations resulting from emissions from up to 50 (total) point and area sources. Calculations of ground-level pollutant concentrations are made for each frequency designed in an array defined by six stabilities, 16 wind directions and six wind speeds for 112 program-designed receptor sites on a radial	Burt (1977)

Table 7.5—contd.

Model	*Model details*	*Reference*
	grid of variable scale. Empirical dispersion coefficients are used and include adjustments for plume rise and limited mixing. Plume height is adjusted according to terrain elevations and stability classes	
RAM	Gaussian-Plume Multiple-Source Air Quality Algorithm. This short-term gaussian steady-state algorithm estimates concentrations of stable pollutants from urban point and area sources. Hourly meteorological data are used. Hourly concentrations and averages over a number of hours can be estimated. Briggs plume rise is used. Pasquill–Gifford dispersion equations with dispersion parameters thought to be valid for urban areas are used. Concentrations from area sources are determined using the method of Hanna, i.e. sources directly upwind are considered representative of area source emissions affecting the receptor. Special features include determination of receptor locations downwind of significant sources and determination of locations of uniformly spaced receptors to ensure good area coverage with a minimum number of receptors	Turner and Novak (1978)

[a] This network, through the EPA's Environmental Applications Branch, makes available air quality dispersion models. In addition to having executable programs on EPA's UNIVAC 1110 at Research Triangle Park, NC, a tape (UNAMAP—version 3) with FORTRAN source programs has been deposited with the National Technical Information Service, US Dept of Commerce, Springfield, VA 22161. The tape accession number is PB-277-193. The above summaries and references were taken by Turner (1979) from the 11 dispersion models contained on the tape.

corporates data on precipitation frequency, intensity, duration and surface area covered. The soil chemistry submodel CHEM estimates the amount of indigenous plus deposited heavy metal in the adsorbed solid and solution phases and the amount of heavy metal moved with eroded soil. The erosion submodel EROS was based on the Universal Soil Loss Equation USLE (Soil Conservation Service, 1976; Wischmeier, 1976) in which the soil loss per unit area per time (A) is given as

$$A = RLSKCP$$

where R is a factor for rainfall, L is the slope length, S is slope

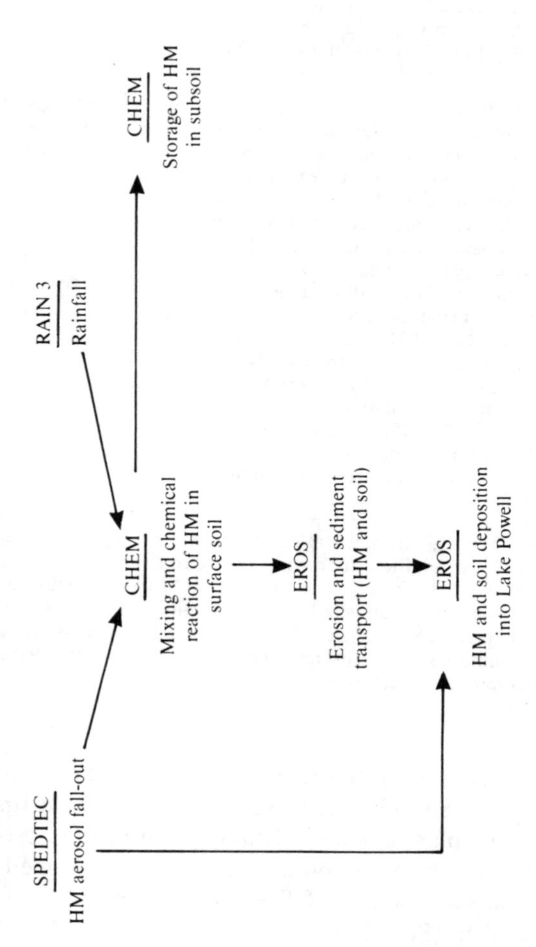

Fig. 7.8. Simplified flow diagram for THOM (first generation model for 'heavy metal' (HM) transport in semiarid and arid environments). Reproduced from Wagnet *et al.* (1978).

steepness, K is slope erodibility, C is cropping, vegetation and management practice and P is conservation practice.

Current knowledge of the use of elemental tracers to identify the source of pollution aerosols has been reviewed by Lowenthal *et al.* (1988) and Lowenthal and Rahn (1988). Seven elements, viz. arsenic, selenium, antimony, zinc, indium, noncrustal manganese and noncrustal vanadium, were measured by instrumental neutron activation in order to provide regional signatures. The chemistry of individual aerosol particles from Chandler, Arizona, an arid urban environment was reported by Anderson *et al.* (1988). Local sources of toxic organic contaminants in the atmosphere have been characterized by pattern recognition techniques (Edgerton & Holdren, 1987), and the atmospheric transport of toxaphene to Lake Michigan has been estimated by Rice *et al.* (1986). Modelling data indicated a toxaphene corridor associated with southerly winds.

Chapter 8

Conclusions

Prediction of the environmental fate of chemicals is a current reality, and is an integral part of evaluations of hazard to the environment and living organisms. However, it is clear that bounds need to be set on any request for environmental fate prediction, since the level of effort and knowledge required is directly related to the specificity of prediction. Thus, a generalized pattern of likely chemical distribution after environmental equilibrium has developed may be predicted with limited resources. Conversely, a demand for information on precise partitioning in a nonequilibrium, and specifically defined situation, will require a commensurate level of input data with the implication of extensive financial cost.

Therefore, as has been suggested by the chapter structure of this book, what is required is a systematic sequence of investigations in order that resources are best utilized. It has been argued that the fundamental physicochemical properties of a chemical should provide the first line of enquiry. These themselves may be amenable to theoretical prediction, using information on molecular structure and chemical bonding. The application of molecular connectivity indices in this context has received particular attention and is commended as a valuable approach. It is apparent that the patterns of binding and behaviour of homologous series of organic chemicals are more predictable than are the behaviour of inorganic compounds and metal ions in natural systems.

Partition coefficient has been demonstrated to be one of the most useful of the molar properties of chemicals that may provide a guide as to behaviour in a wide variety of receptor media. The octanol–water partition coefficient (K_{ow}) mimics the resistance of the bimolecular lipid layer of living cells to chemical exchange and so this has become of particular interest when considering bioaccumulation potential and possible toxicity, serving as a measure of the likely distribution of a substance between lipophilic and water phases.

In conjunction with acquiring information about the fundamental properties of a given substance, it is desirable that an attempt be made to anticipate the possible use to which that substance will be put. Related information would be potential production figures, the means of transporting the substance throughout the distribution network, and the mode of application. For example, aerial spraying of a forest zone near water is more likely to result in contamination of the aquatic phase than surface run-off alone. The proximity of human settlements to a utility which, under worst-case scenarios can emit lethal volatile substances, gives rise to potential for disaster.

Manufacturing processes themselves should be scrutinized to see if particular waste products are inevitable without control measures. Industries such as that of petrochemicals, although involved in the manufacture of countless substances, are never the less dependent on a surprisingly limited number of basic materials or precursor substances. Processing such precursors through particular chemical reactions can lead to the production of intermediates with predictable properties and toxicities.

The controlled disposal of potentially hazardous waste is a potential route to environmental contamination. Thus, fate prediction should also take into consideration containment facilities, leaching potential, biogenic gas emission and degradation rates. Illegal disposal or use in contravention of existing regulations will also have to be accounted for in environmental models attempting to quantify total environmental loadings.

Mathematical modelling is now an integral part of chemical fate prediction exercises, and a vast array of computer programs exists. Some of these are extremely complex and require considerable data input, whereas others are relatively simple and have a minimal data requirement. Each has arguments to commend its particular selection, but the critical point to bear in mind is the nature of the question being put. Thus, there is little merit in utilizing an extravagantly elaborate computer program to solve trivial and generalized estimations of environmental partitioning when all that is required is knowledge of the partition coefficient and the proportion of organic chemical in an aqueous medium. Alternatively, if a complex mix of aerosols is being transported over a whole continent with varying weather patterns and geographic relief, then it may be entirely reasonable to use the most sophisticated computer models available in order to predict the aerosol dispersion pattern. Of the theoretical modelling approaches, some of the most useful and widely used are those incorporating the concept of fugacity.

Empirical models draw more heavily on data provided by environmental studies or results obtained from microcosm experiments. Table 8.1 summarizes some environmental processes and properties, and illustrates the likely minimum data requirement for modelling studies of various physical transport or chemical and biological transformation processes.

Table 8.1
Some environmental processes and properties

A	*W*	*S*[a]	*Process*	*Property*[b]
			Physical transport	
A	—	—	Meteorological	Wind velocity
—	W	S	Bio-uptake	Biomass
—	W	S	Sorption	Organic content of soil or sediment, mass loading of aquatic system, particle size
—	W	S	Volatilization	Turbulence, evaporation rate, reaeration coefficient, soil organic content, surface area, depth
—	—	S	Run-off	Precipitation rate
—	—	S	Leaching	Adsorption coefficient
A	—	—	Fall-out	Particulate concentration, wind velocity
			Chemical transformation	
A	W	—	Photolysis	Solar irradiance, absorption spectra, transmissivity of water, of air
A	W	—	Oxidation	Oxidant and retardant concentrations
—	W	—	Hydrolysis	pH, sediment or soil basicity or acidity
—	W	—	Reduction	Oxygen concentration, ferrous ion concentration, complexation state
			Biological	
—	W	—	Biotransformation/ biodegradation	Microorganism population and acclimation level, trophic status, organism uptake and depuration rates

[a] Probable environmental compartment: A, atmospheric; W, aquatic; S, soil.
[b] At a constant temperature.

Although the fate of chemicals in the environment is potentially determined by transfers between all media, it is often necessary to consider separately processes linked with soil, water and air. This is merely a pragmatic device for dealing with the extreme complexity of interactions between chemicals and the medium to which they are initially released. Multimedia models go on to consider interactions between environmental phases.

The soil compartment is divided into vertically distributed zones, each with characteristic features and implications both for the movement of substances between soil particles and for the binding of chemicals to sorbent surfaces. An integral component of the soil zone is groundwater, the contamination of which is a matter of increasing concern worldwide. The proportion and composition of humic matter affects the sorption of chemicals in the upper soil layers.

The aquatic environment poses an equal challenge to fate prediction in terms of complexity. It also represents the largest sink for the ultimate dispersal of chemicals. There are considerable differences between the marine, estuarine and freshwater environments in terms of acidity and the proportion of inorganic ions. Chemicals may bind to particulate matter in the water column or to sediments. Alternatively, a substance may be soluble under normal environmental conditions. Of particular importance is volatilization from the surface of water bodies, and the impact of certain frequency ranges of solar radiation on chemical transformation and degradation.

The biodegradation of chemicals is discussed in the chapter on the aquatic environment in recognition of the water requirement of living organisms. However, microbial and other biotic activity is of importance to fate prediction in the soil and atmosphere, and represents a vital component in the global cycling of chemicals and their breakdown products.

The fate of chemicals in the atmosphere depends on physical dispersion, particle size, volatility, transformation related to chemical interaction and solar radiation and meteorology. The atmosphere itself should not be considered to be a simple homogeneous medium; it is separated into different zones with increasing altitude, each with its own transformation and dispersal potential. Physical and chemical interactions within clouds are also of great importance. Numerous atmospheric modelling computer programs exist and have been discussed.

It is concluded that facets of the prediction of the environmental fate of chemicals are already well established in numerous scientific disciplines. It is apparent that no one approach is likely to characterize chemical behaviour adequately, rather, the investigator needs sufficient information to be able to select data or models appropriate to specific tasks. Within such bounds there are now excellent tools available for the prediction of biotoxicity and dispersal in soils, water and air. The probability that models will prove reliable is related to our understanding of the environment being simulated and our knowledge of the interacting processes playing upon the substances under consideration.

References and Bibliography

Adams, R. T. and Kurisu, F. M. (1976). *Simulation of pesticide movement on small agricultural watersheds*. Environmental Research Lab., US EPA, Athens, GA, Report No. PB-25933.

Adams, V. D., Watts, R. J. and Pitts, M. E. (1983). (Organics). *J. Wat. Pollut. Contr. Fed.* **55,** 577–98.

Adams, V. D., Watts, R. J. and Pitts, M. E. (1984). (Organics). *J. Wat. Pollut. Contr. Fed.* **56,** 522–44.

Adams, V. D., Watts, R. J. and Pitts, M. E. (1985). (Organics). *J. Wat. Pollut. Contr. Fed.* **57,** 463–93.

Addison, R. F. (1983). PCB replacements in dielectric fluids, *Environ. Sci. Technol.* **17,** 486A–94A.

ADL (1981). *Prepared revisions to MITRE model.* Arthur D. Little Inc., Cambridge, MA.

ADOM/TADAP (1984). *ADOM/TADAP Model Development Program.* Environmental Research and Technology, Concord, MA, Vols 1–7.

ADOM/TADAP (1985). *ADOM/TADAP Model Development Program.* Environmental Research and Technology, Concord, MA, Vol. 8.

Aelion, C. M., Dobbins, C. D. and Pfaender, F. K. (1989). Adaptation of aquifer microbial communities to the biodegradation of xenobiotic compounds: Influence of substrate concentration and preexposure, *Environ. Toxicol. Chem.* **8,** 75–86.

Alexander, M. (1965). Biodegradation: Problems of molecular recalcitrance and microbial fallibility, *Adv. Appl. Microbiol.* **7,** 35–80.

Alexander, M. (1971). *Microbial ecology*. John Wiley and Sons, New York.

Alexander, M. (1973). Biotechnology report: Nonbiodegradable and other recalcitrant molecules, *Biotech. Bioeng.* **15,** 611–47.

Alexander, M. (1978). Biodegradation of toxic chemicals in water and soil. In: *Proc. 176th Natl. Meeting, Miami Beach, FL,* September 1978, Vol. 93, Am. Chem. Soc., Div. Environ. Chem.

Alexander, M. (1985). Biodegradation of organic chemicals, *Environ. Sci. Technol.* **18,** 106–11.

Alexander, M. and Aleem, M. I. H. (1961). Effect of chemical structure on microbial decomposition of aromatic herbicides, *J. Agric. Food Chem.* **9,** 44–7.

Alexander, M. and Lustigman, B. K. (1966). Effect of chemical structure on microbial degradation of substituted benzenes, *J. Agric. Food Chem.* **14,** 410.

Ali, D. A., Callely, A. G. and Hayes, M. (1962). Ability of a vibrio grown in benzoate to oxidize *p*-fluorobenzoate, *Nature* (*London*) **196,** 194–5.

Altenburg, K. (1961). *Kolloid-Z.* **178,** 112.

Altenburg, K. (1966). Die Abhangigkeit der Siedetemperatur isomer Kohlenwasserstoffe von de Form der Molekule. I: Theoretische Grundlagen und die Anwendung auf isomer acylische Alkane. II: Die Siedetemperaturen cyclischer Alkane und die Deutung einiger empirische Regeln zur Konfigurationsanalyse, *Brennst.-Chem.* **47,** 100–7, 331–6.

Ambrose, R. B., Jr, Hill, S. I. and Mulkey, L. A. (1983). *User's Manual for the Toxic Chemical Transport and Fate Model* (*TOXIWASP*) *Version 1.* EPA-600/3-83-005, Athens, GA, 95 pp.

Amidon, G. L. and Anik, S. T. (1976). Comparison of several molecular topological indexes with molecular surface area in aqueous solubility estimation, *J. Pharm. Sci.* **65,** 801–6 (see also Yalkowsky *et al.* (1976)).

Anderson, J. R., Aggett, F. J., Buseck, P. R., Germani, M. S. and Shattuck, T. W. (1988). Chemistry of individual aerosol particles from Chandler, Arizona, an arid urban environment, *Environ. Sci. Technol.* **22,** 811–18.

Anon (1984). *Chem. Eng. News,* Dec. 10, 6–8.

Appel, C. A. and Bredehoeft, J. D. (1978). *Status of ground water modeling in the U.S. Geological Survey.* US Department of the Interior, Washington, DC.

ApSimon, H. and Goddard, A. J. H. (1976). Modeling the atmospheric dispersal of radioactive pollutants beyond the first few hours of travel, *Proc. 7th Int. Tech. Meeting on Air Pollut. Modelling and its Applications,* NATO/CCMS, Airlie, VA, 914–30.

ApSimon, H., Goddard, A. J. H. and Wrigley, J. (1980). Estimating the possible transfrontier consequences of accidental releases: the MESOS model for long range atmospheric dispersal, *Seminar on radioactive releases and their dispersion in the atmosphere following a hypothetical reactor accident,* Riso, Denmark, April 1980, CEC, Luxembourg, 819–42.

Arbuckle, W. B. (1986). Comment on 'Henry's Law Constants for the polychlorinated biphenyls', *Environ. Sci. Technol.* **20,** 527.

Arcos, J. C. (1987). Structure–activity relationships—Criteria for predicting the carcinogenic activity of chemical compounds, *Environ. Sci. Technol.* **21,** 743–5.

Ashbrook, P. C. and Reinhardt, P. A. (1985). Hazardous wastes in academia, *Environ. Sci. Technol.* **19,** 1150–5.

Atkinson, R., Aschmann, S. M. and Winer, A. M. (1987). Kinetics of the reactions of NO_3 radicals with a series of aromatic compounds, *Environ. Sci. Technol.* **21,** 1123–6.

Autian, J. (1975). Structure–toxicity relationships of acrylic monomers, *Environ. Health. Perspect.* **11,** 141–52.

Bachmat, Y., Bredehoeft, J., Andrews, B., Holtz, D. and Sebastian, S. (1980). Ground water management; the use of numerical models, *Water Resources Monograph No. 5,* Am. Geophys. Union, Washington, DC, 127 pp.

Banerjee, S. (1984). Solubility of organic mixtures in water, *Environ. Sci. Technol.* **18,** 587–91.

Banerjee, S. (1985). Calculation of water solubility of organic compounds with UNIFAC-derived parameters, *Environ. Sci. Technol.* **19,** 369–70.

Banerjee, S. and Howard, P. H. (1988). Improved estimation of solubility and partitioning through correction of UNIFAC-derived activity coefficients, *Environ. Sci. Technol.* **22,** 839–41.

Banerjee, S., Yalkowsky, S. H. and Valvani, S. C. (1980). Water solubility and octanol/water partition coefficients of organics. Limitation of the solubility–partition coefficient correlation. *Environ. Sci. Technol.* **14,** 1227–9.

Banerjee, S., Howard, P. H., Rosenberg, A. M., Dombrowski, A. E., Sikka, H. and Tullis, D. L. (1984). Development of a general kinetic model for biodegradation and its application to chlorophenols and related compounds, *Environ. Sci. Technol.* **18,** 416–22.

Banerjee, S., Howard, P. H. and Tullis, D. L. (1985). (Response to comment by Beltrane, P. *et al.,* 1985), *Environ. Sci. Technol.* **19,** 374–75.

Banewicz, J., Reed, C. and Levitch, M. (1957). Experimental investigation of the distribution of salicylic acid between cyclohexane and water, *J. Am. Chem. Soc.* **79,** 2693–5.

Barcelona, M. J. and Naymik, T. G. (1984). Dynamics of a fertilizer contaminant plume in groundwater, *Environ. Sci. Technol.* **18,** 257–61.

Barry, T. M. (1987). Integrated environmental management: a pilot study of the pulp and paper industry, *Industry and Environment* **10,** 29–34.

Bauer, A. L. (1987). *Water Resour. Res.* **23,** 1926–38.

Baughman, G. L. and Lassiter, R. R. (1978). Prediction of environmental pollutant concentration. In: *ASTM STP 657,* J. Carins Jr, K. L. Dickson and A. W. Maki (Eds), 35–54.

Baumgartner, D. J. and Callaway, R. J. (1972). State of the art for simulation of pollution problems and controls in estuaries. In: *Marine Pollution and Sea Life,* M. Ruivo (Ed.), FAO/Fishing News (Books) Ltd, 625 pp. From: *FAO Tech. Conf. on Mar. Pollut. and its effects on living resources and fishery,* Rome, December 1970, 140–6.

Baxter, R. M. and Sutherland, D. A. (1984). Biochemical and photochemical processes in the degradation of chlorinated biphenyls, *Environ. Sci. Technol.* **18,** 608–10.

Bear, J. (1979). *Hydraulics of Ground Water.* McGraw-Hill, New York.

Bell, C. R. (1960). A soil achromobacter which degrades 2,4-dichlorophenoxyacetic acid, *Can. J. Microbiol.* **6,** 325–337.

Beltrame, P., Beltrame, P. L. and Carniti, P. (1985). Comment on 'Development of a general kinetic model for biodegradation and its application to chlorophenols and related compounds', *Environ. Sci. Technol.* **19,** 374.

Bennett, B. G. (1982). The exposure commitment method for pollutant exposure evaluation, *Ecotoxicol. Environ. Safety,* **6,** 363–8.

Bennett, D. J., Dence, C. W., Kung, F. L., Luner, P. and Ota, M. (1971). Mechanism of color removal in treatment of spent bleaching liquors with lime, *Tappi* **54,** 2019.

Berthelot and Jungfleisch (1872). Sur les lois qui president au partage d'un corps entre deux dissolvants (experiences), *Ann. Chim. Phys.* 4[e] Serie, Tome XXVI, 396–407.

Betterton, E. A. and Hoffman, M. R. (1988). Henry's Law Constants of some environmentally important aldehydes, *Environ. Sci. Technol.* **22,** 1415–18.

Bhumralker, C. M., Mancuso, R. L., Wolf, D. E., Johnson, W. B. and Pankrath, J. (1981). Regional air pollution model for calculating short-term (daily) patterns and transfrontier exchanges of airborne sulfur in Europe, *Tellus,* **33,** 142–61.

Bidleman, T. F. (1988). Atmospheric processes, *Environ. Sci. Technol.* **22,** 361–7.

Blair, E. H. (1981). A framework of consideration for setting priorities for the testing of chemicals. Paper presented to the OECD, Berlin, FRG. (Cited in Neely, 1982.)

Blau, G. E. and Neely, W. B. (1983). What constitutes an adequate model for predicting the behaviour of pesticides in the environment? In: *Residue Reviews 85,* F. A. Gunter and J. D. Gunter (Eds), Springer-Verlag, Berlin, 293–307.

Boatman, R. J., Cunningham, S. L. and Ziegler, D. A. (1986). A method for measuring the biodegradation of organic chemicals, *Environ. Toxicol. Chem.* **5,** 233–43.

Bokova, E. N. (1954). Oxidation of ethane and propane by Mycobacterium. *Mikrobiologiya* **23,** 15–21.

Bolin, B. and Persson, C. (1975). Regional dispersion and deposition of atmospheric pollutants with particular application to sulfur pollution over Western Europe, *Tellus* **27,** 281–310.

Bolin, B. and Rodhe, H. (1973). A note on the concepts of age distribution and transit time in natural reservoirs, *Tellus* **25,** 58–62.

Bomberger, D. C., Gwinn, J. L., Mabey, W. R., Tuse, D. & Chou, T. W. (1983). Environmental fate and transport at the terrestrial–atmospheric interface. In: *Fate of Chemicals in the Environment,* R. L. Swann and A. Eschenroeder (Eds), ACS Symp. Ser. 225 (Chapter 10), 197–214.

Bonazountas, M. (1983). Soil and ground water fate modeling. In: *Fate of Chemicals in the Environment,* R. L. Swann and A. Eschenroeder (Eds), ACS Symp. Ser. 225 (Chapter 3), 41–65.

Bonazountas, M. and Fiksel, J. (1982). *Enviro: Environmental Mathematical Pollutant Fate Modeling Handbook/Catalogue.* EPA Contract No. 68-01-5146, Arthur D. Little Inc., Cambridge, MA.

Bonazountas, M. & Wagner, J. (1981). *SESOIL: A seasonal soil compartment mode.* Office of Toxic Substances, US EPA, Washington, DC.

Bondi, A. (1964). Van der Waals Volumes and Radii, *J. Phys. Chem.* **68,** 441–51.

Bostater, C. R. Jr, Ambrose, R. B., Jr and Bell, B. (1981). Modeling the fate and transport of chemicals in estuaries: current approaches and future needs. In: *Aquatic Toxicology and Hazard Assessment: 4th Conf.,* ASTM

STP 737, D. R. Branson and K. L. Dickson (Eds), Am. Soc. for Testing and Materials, 1981, 72–90.

Bowen, H. J. M. (1975). Residence times of heavy metals in the environment. In: *Int. Conf. on Heavy Metals in the Environment,* Canada, 1975, Vol. 1, Symp. Proc., 1–19.

Brandstrom, A. (1963). A rapid method for the determination of distribution coefficients of bases for biological purposes, *Acta Chem. Scand.* **17,** 1218–24.

Branson, D. R. (1978). Predicting the fate of chemicals in the aquatic environment from laboratory data. In: *Estimating the Hazard of Chemical Substances to Aquatic Life,* J. J. Cairns, K. L. Dickson and A. W. Maki (Eds), Am. Soc. for Testing and Materials, 1978, ASTM STP 657, 55–70.

Brooke, D. N., Dobbs, A. J. and Williams, N. (1986). Octanol:water partition coefficients (*P*): Measurement, estimation, and interpretation, particularly for chemicals with $P > 10^{51}$, *Ecotoxicol. Environ. Safety* **11,** 251–60.

Browman, M. G., Patterson, M. R. and Sworski, T. J. (1982). *Formulations of the physiochemical processes in the ORNL Unified Transport Model for Toxicants (UTM-TOX), Interim Report.* ORNL/TM-8013, Oak Ridge Nat. Lab., TN, 46 pp.

Brown, S. L. and Bomberger, D. C. (1983). Release of chemicals into the environment. In: *Fate of Chemicals in the Environment—Compartmental and Multimedia Models for Predictions,* R. L. Swann and A. Eschenroeder (Eds), ACS Symp. Ser. **225,** 3–21.

Brown, S. L., Holt, B. R. and McCaleb, K. E. (1978). *Systems for rapid ranking of environmental pollutants. Selection of subjects for scientific and technical assessment reports.* EPA-600/5/-78-012 (June 1978), 143, 199, 200.

Brubaker, K. L., Brown, P. and Cirillo, R. R. (1977). *Addendum to User's Guide for Climatological Dispersion Model.* Prepared by Argonne Nat. Lab. for US EPA, Research Triangle Park, NC, EPA-450/3-77-015, NTIS PB 274-040.

Bruggeman, W. A., van der Steven, J. and Hutzinger, J. (1982). Reversed-phase thin-layer chromatography of polynuclear aromatic hydrocarbons and chlorinated biphenyls—relationship with hydrophobicity as measured by aqueous solubility and octanol water partition-coefficient, *J. Chromatogr.* **238,** 335–46.

Buat-Menard, P. E. (1983). Particle geochemistry in the atmosphere and oceans. In: *Air–Sea Exchange of Gases and Particles* (Chapter 10), P. S. Liss and W. G. N. Slinn (Eds), Dordrecht, Holland: Reidel.

Buat-Menard, P. E. (1986). *The Role of Air–Sea Exchange in Geochemical Cycling.* NATO-ASI Series 185, Reidel, Boston, MA.

Budiansky, S. (1980). Dispersion modelling, *Environ. Sci. Technol.* **14,** 370–4.

Burger, K., MacRae, I. C. and Alexander, M. (1962). Decomposition of phenoxyalkyl carboxylic acids, *Soil Sci. Soc. Am. Proc.* **26,** 243.

Burkhard, L. P., Andrew, A. W. and Armstrong, D. E. (1985*a*). Estimation of vapor pressures for polychlorinated biphenyls: A comparison of eleven predictive methods, *Environ. Sci. Technol.* **19,** 500–7.

Burkhard, L. P., Armstrong, D. E. and Andrew, A. W. (1985*b*). Henry's Law

Constants for the polychlorinated biphenyls, *Environ. Sci. Technol.* **19,** 590–6.

Burkhard, L. P., Andrew, A. W., Loux, N. T. and Armstrong, D. E. (1986) (Response to Arbuckle, W. B. (1986)). *Environ. Sci. Technol.* **20,** 527–8.

Burns, L. A. (1983). Fate of chemicals in aquatic systems: process models and computer codes. In: *Fate of Chemicals in the Environment—Compartmental and Multimedia Models for Predictions,* R. L. Swann and A. Eschenroeder (Eds), (Chapter 2), ACS Symp. Ser. 225, 25–40.

Burns, L. A., Cline, D. M. and Lassiter, R. R. (1982). *Exposure Analysis Modelling System (EXAMS): User Manual and System Documentation.* EPA-600/3-82-023, US EPA Environ. Res. Lab., Athens, GA, 443 pp.

Burris, D. R. and MacIntyre, W. G. (1985). Water solubility behaviour of binary hydrocarbon mixtures, *Environ. Toxicol. Chem.* **4,** 371–7.

Burt, E. W. (1977). *Valley Model User's Guide.* US EPA, Research Triangle Park, NC, EPA-450/2-77-018, NTIS PB 274-054.

Busse, A. D. and Zimmerman, J. R. (1973). *User's Guide for the Climatological Dispersion Model.* Environmental Monitoring Ser., US EPA, Research Triangle Park, NC, EPA-RA-73-024, NTIS PB 227-346.

Bysshe, S. E. (1982). Bioconcentration factor in aquatic organisms. (Chapter 5) In: *Handbook of Chemical Property Estimation Methods: Environmental Behaviour of Organic Compounds,* W. J. Lyman, W. F. Reehl and D. H. Rosenblatt (Eds), McGraw-Hill, New York, pp. 5-1, 5-30.

Cady, R. E. and Peckenpaugh, J. M. (1985). *RAQSIM Users Guide: Regional Aquifer Simulation Model.* USGS Water Resources Investigations Report 85-4168, Lincoln, NE (More information available from the Open-file Services Section, US Geological Survey, Box 25425, Federal Center, Denver, CO.)

Cairns, J., Jr, (1982). Predictive and reactive systems for aquatic ecosystem quality control. In: *Scientific Basis of Water-Resource Management,* Geophysics Study Committee, National Res. Council, National Academy Press, Washington, DC, 72–84.

Cairns, J., Jr, Dickson, K. L. and Maki, A. W. (Eds) (1978). *Estimating the hazard of chemical substances to aquatic life.* Sponsored by ASTM Committee D.19 on water. Am. Soc. for Testing and Materials, Special Tech. Publ. 657. 278 pp.

Callahan, M. A. and Slimak, M. W. (1979). *Water related environmental fate of 129 priority pollutants.* EPA-440/4-79-029 (a) and (b).

Cammarata, A., Yau, S. J. and Rogers, K. S. (1971). Structure–activity relationships having a basis in regular solution theory, *J. Med. Chem.* **14,** 1211–12.

Campbell, J. R. and Luthy, R. G. (1985). Prediction of aromatic solute partition coefficients using the UNIFAC group contribution model, *Environ. Sci. Technol.* **19,** 980–5.

Campbell, J. R., Luthy, R. G. and Carrondo, M. J. T. (1983). Measurement and prediction of distribution coefficients for wastewater aromatic solutes, *Environ. Sci. Technol.* **17,** 582–90.

Canale, R. P. and Auer, M. T. (1987). Personal computers and environmental engineering (Part II—Applications), *Environ. Sci. Technol.* **21,** 936–42.

Capel, P. D., Giger, W., Reichert, P. and Wanner, O. (1988). Accidental input of pesticides into the Rhine River, *Environ. Sci. Technol.* **22,** 992–7.

Carlson, R. M., Carlson, R. E. and Kopperman, H. L. (1975). Determination of partition-coefficients by liquid-chromatography, *J. Chromatogr.* **107,** 209–23.

Carmichael, G. R. and Peters, L. K. (1979). Numerical simulation of the regional transport of SO_2 and sulfate in the eastern United States. *Proc. 4th Symp. Turbulence Diffusion and Air Pollution,* Reno, NV, 1979, Am. Met. Soc., 337–44.

Carmichael, G. R. and Peters, L. K. (1980). The transport, chemical transformation and removal of SO_2 and sulfate in the Eastern United States. In: *Atmospheric Pollution,* M. Benarie (Ed.), Elsevier, Amsterdam, 31–6.

Carmichael, G. R., Peters, L. K. and Kitada, T. (1986). A 2nd generation model for regional-scale transport chemistry deposition, *Atmos. Environ.* **20,** 173–88.

Carsel, R. F. (1980). *Pesticide Runoff Simulator User's Manual.* Computer Sciences Corporation. (Carsel was based at EPA, Athens, GA).

Carter, C. W. and Suffet, I. H. (1983). Interactions between dissolved humic and fulvic acids and pollutants in aquatic environments. In: *Fate of Chemicals in the Environment,* R. L. Swann and A. Eschenroeder (Eds), ACS Symp. Ser. 225 (Chapter 11), 215–29.

Cayley, E. (1874). *Phil. Mag.* **67,** 444.

Chambers, C. W. and Kabler, P. W. (1964). Biodegradability of phenols as related to chemical structure, *Dev. Ind. Microbiol.* **5,** 85.

Chawla, R. C. and Varma, M. M. (1981/82). Pollutant transfer between air, water and soil: Criteria for comprehensive pollution control strategy, *J. Environ. Syst.* **11,** 363–74.

Chiou, C. T. (1985). Partition coefficients of organic compounds in lipid–water systems and correlations with fish bioconcentration factors, *Environ. Sci. Technol.* **19,** 57–62.

Chiou, C. T. and Freed, V. H. (1977). *Chemodynamic studies on benchmark industrial chemicals.* NSF/RA-770286, NTIS PB 274263.

Chiou, C. T. and Freed, V. H. (1979). Evaporation rates from single-component and multicomponent systems. Preprints of papers represented at *177th Natl. Meeting,* Honolulu, Hawaii, April 1–6, 1979. American Chem. Soc., Div. Environ. Chem., Vol. 19, No. 1.

Chiou, C. T., Malcolm, R. L., Brinton, T. I. and Kile, D. E. (1986). Water solubility enhancement of some organic pollutants and pesticides by dissolved humic and fulvic acids, *Environ. Sci. Technol.* **20,** 502–8.

Christensen, E. R. and Nyholm, N. (1984). Ecotoxicological assays with algae: Weibull dose–response curves, *Environ. Sci. Technol.* **18,** 714–18.

Christensen, O. and Prahm, L. P. (1976). *A pseudospectral model for dispersion of atmospheric pollutants. 1. Numerical tests.* Danish Meteorological Institute, Air Pollution Section, Lyngbyvej Loo, DK-2100, 38 pp.

Christian, S. D., Affsprung, H. E., Johnson, J. R. and Worley, J. D. (1963). Control and measurement of water activity, *J. Chem. Educ.* **40,** 419–21.

Chudyk, W. (1989). Field screening of hazardous waste sites, *Environ. Sci. Technol.* **23,** 503, 504–7.

Coates, J. T. (1984). PhD Dissertation, Clemson University, Clemson, SC.

Coates, M., Connell, I. W. and Barron, D. M. (1985). Aqueous solubility and octan-1-ol to water partition coefficients of aliphatic hydrocarbons, *Environ. Sci. Technol.* **19,** 628–32.

Cohen, M. L. and Steinmetz, W. D. (1986). Foliar wash-off of pesticides by rainfall, *Environ. Sci. Technol.* **20,** 521–3.

Collander, R. (1951). The partition of organic compounds between higher alcohols and water, *Acta Chem. Scand.* **5,** 774–80.

Connell, D. W. and Schüürmann, G. (1988). Evaluation of various molecular parameters as predictors of bioconcentration in fish, *Ecotoxicol. Environ. Safety* **15,** 324–5.

Connolly, J. P. (1982). *WASTOX preliminary estuary and stream version documentation.* US EPA, Environ. Res. Lab., Gulf Breeze, FL, 96 pp. (Draft.)

Conover, W. J. (1971). *Practical Nonparametric Statistics.* John Wiley and Sons, New York, 462 pp.

Cox, J. D. and Pilcher, G. (1970). *Thermochemistry of Organic and Organometallic Compounds.* Academic Press, New York.

Craig, L. C. and Craig, D. (1950). In: *Technique of Organic Chemistry, Vol. III, Part I,* A. Weissberger (Ed.), Interscience, New York, 171 pp.

Craig, L. C., Hogeboom, G. H., Carpenter, F. H. and Du Vigneaud, V. (1947). Separation and characterization of some penicillins by the method of counter-current distribution, *J. Biol. Chem.* **168,** 665–86.

Craig, P. J. (1983). Organometallic Compounds in the Environment. In: *Pollution: Causes, Effects and Control,* R. M. Harrison (Ed.), Roy. Soc. Chem. Special Publ. No. 44 (Chapter 16), 278–322.

Crist, R. H., Oberholser, K., Schwartz, D., Marzoff, J., Ryder, D. and Crist, D. R. (1988). Interactions of metals and protons with algae, *Environ. Sci. Technol.* **22,** 755–60.

Crittenden, J. C., Luft, P., Hand, D. W., Oravitz, J. L., Loper, S. W. and Ari, M. (1985). Prediction of multicomponent adsorption equilibria using ideal adsorbed solution theory, *Environ. Sci. Technol.* **19,** 1037–43.

Crum Brown, A. and Fraser, T. R. (1867/68). On the connection between chemical constitution and physiological action. Part I—on the physiological action of the salts of the ammonium bases, derived from Strychnia, Brucia, Thebaia, Codeia, Morphia, and Nicotia. *Trans. R. Soc. Edinburgh,* **25,** for the session 1867/68, 151–203.

Crynes, B. L. (1981). In: *Chemistry of Coal Utilization,* 2nd Supplementary Volume, M. A. Elliot (Ed.), Wiley-Interscience, New York, 1991–2070.

Cumberland, R. F. (1982). The control of hazardous chemical spills in the United Kingdom, *J. Hazard. Mat.* **6,** 277–87.

Curds, C. R. (1974). Computer simulations of some complex microbial food chains, *Water Res.* **8,** 769–80.

Dabberdt, W. F., Ludwig, F. L. and Johnson, W. B. Jr (1973). Validation and applications of an urban diffusion model for vehicular pollutants, *Atmos. Environ.* **7,** 603.

Daisey, J. M. (1980). Organic compounds in urban-aerosols, *Ann. N.Y. Acad. Sci.* **338,** 50–69.

Davidar, D. (1985). Beyond Bhopal: The toxic waste hazard in India, *Ambio* **14,** 112–16.

Davies, C. W. (1962). *Ion Association.* Butterworths, London.

Davies, M. and Griffiths, D. (1954). *Z. Phys. Chem.* (Frankfurt am Main) **2,** 353.

Davies, M. and Griffiths, D. M. L. (1955). Intramolecular hydrogen bonds and the association and solubilities of substituted benzoic acids, *J. Chem. Soc.* 132–7.

Davies, M., Jones, P., Patnaik, D. and Moelwyn-Hughes, E. A. (1951). The distribution of the lower fatty acids between water and a variety of solvents, *J. Chem. Soc.* 1249–52.

Davies, R. P. and Dobbs, A. J. (1984). The prediction of bioconcentration in fish, *Water Res.* **18,** 1253–62.

Davies-Colley, R. J., Nelson, P. O. and Williamson, K. J. (1984). Copper and cadmium uptake by estuarine sedimentary phases, *Environ. Sci. Technol.* **18,** 491–9.

Dean, R. B. (1981). In: *Chemistry in Water Reuse,* Volume 1, W. J. Cooper (Ed.), Ann Arbor Science Publications, Ann Arbor, MI.

Deegan, J., Jr (1987*a*). Looking back at Love Canal, *Environ. Sci. Technol.* **21,** 328–31.

Deegan, J., Jr (1987*b*). Looking back at Love Canal (Results and conclusion of the EPA's investigation), *Environ. Sci. Technol.* **21**(5), 421–6.

Deininger, R. L. (1987). The survival of Father Rhine, *J. Am. Water Works Assoc.* **79,** 78–83.

Dennis, R. L. (1980). An air-pollution dispersion model for long-range policy analysis, *Ecol. Modelling* **9,** 121–41.

de Pastrovich, T. L. *et al.* (1979). *CONCAWE's Water Pollution Special Task Force No. 11 (1979): Protection of ground water from oil pollution.* CONCAWE Report No. 3./79.

De Walle, F. B., Lo, C., Sung, J., Kalman, D., Chian, E. S. K., Giabbai, M. and Ghosai, M. (1982). (Organics). *J. Wat. Pollut. Contr. Fed.* **54,** 555–76.

De Walle, F. B., Norman, D., Sung, J., Chian, E. S. K. and Giabbai, M. (1981). (Organics). *J. Wat. Pollut. Contr. Fed.* **53,** 659–74.

Dias, F. F. and Alexander, M. (1971). The effects of chemical structure on the biodegradability of aliphatic acids and alcohols, *Appl. Microbiol.* **22,** 1114.

Dickson, K. L., Maki, A. W. and Cairns, J., Jr (1982). In: *Modelling the fate of chemicals in the Aquatic Environment,* Ann Arbor Science Publications, Ann Arbor, MI, p. 413.

Dilling, W. L. (1977). Interphase transfer processes. II. Evaporation rates of chloro methanes, ethanes, ethylenes, propanes, and propylenes from dilute aqueous solutions. Comparisons with theoretical predictions, *Environ. Sci. Technol.* **11,** 405–9.

Di Toro, D. M., Fitzpatrick, J. J. and Thomann, R. V. (1981*a*). *Water Quality Analysis Simulation Program (WASP) and Model Verification Program (MVP) . . . Documentation,* EPA/68-01-3872. US EPA, Environ. Res. Lab., Duluth, MN, 135 pp.

Di Toro, D. M., O'Connor, D. J., Thomann, R. V. and St. John, J. P. (1981*b*). *Analysis of Fate of chemicals in receiving waters.* Phase 1, CMA Project ENV-7-W. Hydro Qual., Mahwah, NJ.

Di Toro, D. M., O'Connor, D. J., Thomann, R. V. and St. John, J. P. (1982). In: *Modelling the Fate of Chemicals in the Aquatic Environment,* K. L. Dickson, J. Cairns Jr, and A. W. Maki (Eds), Ann Arbor Science Publications, MI, 165–90.

Di Toro, D. M., Jeris, J. S. and Ciarcia, D. (1985). Diffusion and partitioning of hexachlorobiphenyl in sediments, *Environ. Sci. Technol.* **19,** 1169–76.

Di Toro, D. M., Mahony, J. D., Kirchgraber, P. R., O'Byrne, A. L., Pasquale, L. R. and Piccirilli, D. C. (1986). Effects of nonreversibility, particle concentration and ionic strength on heavy metal sorption, *Environ. Sci. Technol.* **20,** 55–61.

Doherty, F. G. (1983). Interspecies correlations of acute median lethal concentration for four standard testing species. *Environ. Sci. Technol.* **17,** 661–5.

Donigian, A. S., Jr (1983). Model predictions vs. field observations: the model validation/testing process. In: *Fate of Chemicals in the Environment—Compartmental and Multimedia Modes for Predictions,* R. L. Swann and A. Eschenroeder (Eds), ACS Symp. Ser. 225 (Chapter 8), 151–71.

Donovan, R. J. (1983). Pollutant pathways in the atmosphere. In: *Pollution: Causes, Effects and Control,* R. M. Harrison (Ed.), Roy. Soc. Chem., Special Publ. No. 44 (Chapter 9), 176-90.

Doucette, W. J. and Andrew, A. W. (1987). Correlation of octanol/water partition coefficients and total molecular surface area for highly hydrophobic aromatic compounds, *Environ. Sci. Technol.* **21,** 821–24.

Doury, A. (1982). Operational calculation aids for atmospheric dispersion, *Sci. Tot. Environ.* **25,** 3–17.

Dunn, W. J. III and Wold, S. (1980). Relationships between chemical structure and biological activity modelled by SIMCA pattern recognition, *Bioorganic Chem.* **9,** 505–23.

Dunnivant, F. M., Coates, J. T. and Elzerman, A. W. (1988). Experimentally determined Henry's Law Constants for 17 polychlorinated congeners, *Environ. Sci. Technol.* **22,** 448–53.

Durham, J. (1982). Atmospheric chemistry. In: *United States–Canada Memorandum of Intent on Transboundary Air Pollution,* Atmospheric Sciences and Analysis Work Group 2, Report OSE 81-00164, prepared for Atmospheric Environment Service by Environmental Applications Group Ltd, Toronto, Canada, 71 pp.

Dyer, J. R. (1965). *Applications of absorption spectroscopy of organic compounds.* Prentice-Hall, Englewood Cliffs, New Jersey, 4–21.

Dzombak, D. A., Fish, W. and Morel, F. M. M. (1986). Metal–humate interactions. 1. Discrete ligand and continuous distribution models, *Environ. Sci. Technol.* **20,** 669–75.

Eaton, J. G. *et al.* (Eds) (1980). Aquatic toxicology. *Proc. 3rd. Ann. Symp. on Aq. Toxicol.,* sponsored by ASTM committee E-35 on pesticides, New Orleans, LA, October 1978, ASTM STP 707, 405 pp.

ECETOC (1986). *Structure–activity relationships in toxicology and ecotoxicology: An assessment.* European Chemical Industry Ecology and Toxicology Centre, Monograph No. 8, Brussels, Belgium, 88 pp.

Eckel, W., Trees, D., Viar and Company, Kovell, S. (EPA) (1985).

Contaminants found at Superfund Sites, Presentation at the National Conference on Hazardous Wastes and Environmental Emergencies, Cincinnati, Ohio, May 1985. (cited by Josephson, J. (1985)).

Edgerton, S. A. and Holdren, M. W. (1987). Use of pattern recognition techniques to characterize local sources of toxic organics in the atmosphere, *Environ. Sci. Technol.* **21,** 1102–7.

Eliassen, A. (1976). *The trajectory model: A technical description.* Norwegian Institute for Air Research, PO Box 115, N-2007 Kjeller, Norway, LRTAP No. 1/76, 14 pp.

Eliassen, A. (1978). The OECD study of long-range transport by air pollutants: long-range transport modelling, *Atmos. Environ.* **12,** 479–87.

Eliassen, A. and Saltbones, J. (1975). Decay and transformation rates of SO_2, as estimated from emission data, trajectories and measured air concentrations, *Atmos. Environ.* **9,** 425–9.

Ellgehausen, H., Guth, J. A. and Esser, H. O. (1980). Factors determining the bioaccumulation potential of pesticides in the individual compartments of aquatic food chains, *Ecotoxicol. Environ. Safety* **4,** 134–57.

Ellgehausen, H., D'Hondt, C. and Fuerer, R. (1981). Reversed-phase chromatography as a general method for determining octan-1-ol/water partition coefficients, *Pest. Sci.* **12,** 219–27.

Ellis, J. H. (1988). Acid rain control strategies, *Environ. Sci. Technol.* **22,** 1248–55.

EMEP (1980). *Summary report of the Western Meteorological Synthesizing Centre for the first phase of EMEP.*

Enfield, C. G., Carsel, R. F., Cohen, S. Z., Phan, T. and Walters, D. M. (1980). *Approximating pollutant transport to ground water.* US EPA, RSKERL, Ada, OK. (Unpublished.)

England, R. L., Ekblad, K. J. and Baca, R. G. (1985). *MAGNUM-2D Users Guide: RHO-BW-CR-143.* Rockwell International, Richland, WA.

Englande, A. J. and Reimers, R. S., III (1980). Hazardous industrial waste characteristics: An overview, *Proc. Conf. Hazardous and Toxic Waste Management,* 545–57.

EPA (1979). *Technical Environmental Impacts of Various Approaches for Regulating Small Volume Hazardous Waste Generators,* Volume 1, Office of Solid Waste and Emergency Responses, Environmental Protection Agency, Washington, DC.

EPA (1982). *Environmental Monitoring at Love Canal,* volumes I–III, EPA 600/4-82-030 a-c, NTIS PB82-237330, Office of Research and Development, US Environmental Protection Agency, Washington, DC.

Ephraim, J. and Marinsky, J. A. (1986). A unified physicochemical description of the protonation and metal ion complexation equilibria of natural organic acids (humic and fulvic acids). 3. Influence of polyelectrolyte properties and functional heterogeneity on the copper ion binding equilibria in an Armadale Horizons Bh fulvic acid sample, *Environ. Sci. Technol.* **20,** 367–70.

Ephraim, J., Alegret, S., Mathuthu, A., Bicking, M., Malcolm, R. L. and Marinsky, J. A. (1986). A unified physicochemical description of the protonation and metal ion complexation equilibria of natural organic acids (humic and fulvic acids). 2. Influence of polyelectrolyte properties and functional group heterogeneity on the protonation equilibria of fulvic acid, *Environ. Sci. Technol.* **20,** 354–66.

Epstein, C. B. (1988). A method for the estimation of historical sulfate concentrations in natural freshwaters, *Environ. Sci. Technol.* **22,** 1460–3.

Eschenroeder, A. (1975). *An assessment of models for predicting air quality.* ERT Document ERTW-75-03, Environmental Research and Technol., Inc., Santa Barbara, CA, 83 pp.

Eschenroeder, A. (1981). Multimedia modeling of the fate of environmental chemicals. In: *Aquatic Toxicology and Hazard Assessment: 4th Conf.*, D. R. Branson and K. L. Dickson (Eds), ASTM STP 737, American Society for Testing and Materials, 31–42.

Esser, H. O. and Moser, P. (1982). An appraisal of problems related to the measurement and evaluation of bioaccumulation, *Ecotox. Environ. Safety* **6,** 131–48.

Falkengren-Grerup, U. (1987). Long-term change in pH of forest soils in southern Sweden, *Environ. Pollut.* **43,** 79–90.

Farman, J. C., Gardiner, B. G. and Shanklin, J. D. (1985). Large losses of total ozone reveal seasonal ClO_x/NO_x interaction, *Nature* (*Lond.*) **315,** 207–10.

Faust, B. C. and Higné, J. (1987). Sensitized photooxidation of phenols by fulvic acid and in natural waters, *Environ. Sci. Technol.* **21,** 957–64.

Federal Register (1982). The Hazard Ranking System, *Fed. Regist.* **47,** 31186.

Felmy, A. R., Brown, S. M., Onishi, Y., Yabusaki, S. B., Argo, R. S., Girvin, D. C. and Jenne, E. A. (1984). *Modeling the Transport, Speciation, and Fate of Heavy Metals, Aquatic Systems,* Project Summary, EPA-600/S3-84-033, April 1984, US EPA, Environmental Research Laboratory, Athens, Galveston, 4 pp. (EPA Project Officer: R. B. Ambrose).

Fenchel, T. M. and Jorgensen, B. B. (1977). Detritus food chains of aquatic ecosystems: The role of bacteria. In: *Advances in Microbial Ecology,* M. Alexander (Ed.), Plenum Press, New York.

Fendinger, N. J. and Glotfelty, D. E. (1988). A laboratory method for the experimental determination of air–water Henry's Law Constants for several pesticides, *Environ. Sci. Technol.* **22,** 1289–93.

Figge, K., Klahn, J. and Koch, J. (1986). Kinetic distribution model for chemicals based on results from a standard environmental system, *Ecotox. Environ. Health* **11,** 320–38.

Finnerty, W. R., Hawtrey, E. and Kallio, R. E. (1962). Alkane-oxidizing micrococci, *Z. Allg. Mikrobiol.* **2,** 169.

Fish, W., Dzombak, D. A. and Morel, F. M. M. (1986). Metal–humate interactions. 2. Application and comparison of models, *Environ. Sci. Technol.* **20,** 676–83.

Fisher, B. E. A. (1983). A review of the processes and models of long-range transport of air pollutants, *Atmos. Environ.* **17,** 1865–80.

Foote, H. P. (1982). For information: Battelle Pacific Northwest Labs, PO Box 99, Richland, VA. (re: MMT/VTT soil models).

Fredenslund, A., Jones, R. L. and Prausnite, J. M. (1975). Group-concentration estimation of activity-coefficients in nonideal liquid-mixtures, *AIChE J.* **21,** 1086–99.

Freeze, R. A. and Cherry, J. A. (1979). *Ground Water*. Prentice-Hall, Englewood Cliffs, NJ.

Frische, R., Klopffer, W. and Schonborn, W. (1979*a*). *Bewertung von organisch-chemischen Stoffen und Produkten in bezug auf ihr Umweltverhalten—chemische, biologische und wirtschaftliche Aspekte*. Battelle-Institut e.V., Frankfurt, FRG, BF-R-63, 560-1.

Frische, R., Kloppfer, W. and Schonborn, W. (1979*b*). *Bewertung von organisch—chemischen Stoffen und Produckten in bezug auf ihr Umweltverhalten-chemische, biologische und wirtschaftliche Aspekte*. Final Report BF-R-63.560-01 of Battelle-Institut e.V., Frankfurt am Main, to Umweltbundesamt, Berlin, May 1979.

Frische, R., Esser, G., Schonborn, W. and Klopfer, W. (1982). Criteria for assessing the environmental behaviour of chemicals: Selection and preliminary quantifications, *Ecotoxicol. Environ. Safety* **6,** 283–93.

Fujita, T., Iwasa, J. and Hansch, C. (1964). A new substituent constant, derived from partition coefficients, *J. Am. Chem. Soc.* **86,** 5175–80.

Gabric, A. J. (1983). Effect of shear and surface boundary on contaminant diffusion in the ocean, *Aust. J. Mar. Freshwater Res.* **34,** 693–706.

Gaffney, J. S., Streit, G. E., Spall, W. D. and Hall, J. H. (1987). Beyond acid rain—Do soluble oxidants and organic toxins interact with SO_2 and NO_x to increase ecosystem effects? *Environ. Sci. Technol.* **21,** 519–24.

Galloway, J. N., Norton, S. A. and Church, M. R. (1983). Freshwater acidification from atmospheric deposition of sulfuric acid, a conceptual model, *Environ. Sci. Technol.* **17,** 541A–5A.

Gamble, D. S. and Langford, C. H. (1988). Complexing equilibria in mixed ligand systems: Tests of theory with computer simulations, *Environ. Sci. Technol.* **22,** 1325–36.

Garland, J. A. (1978). Dry and wet removal of sulphur from the atmosphere, *Atmos. Environ.* **12,** 349–62.

Garten, C. T., Jr and Trabalka, J. R. (1983). Evaluation of models for predicting terrestrial food chain behaviour of xenobiotics, *Environ. Sci. Technol.* **17,** 590–5.

Gauthier, T. D., Shane, E. C., Guerin, W. F., Seitz, W. R. and Grant, C. L. (1986). Fluorescence quenching method for determining equilibrium constants for polycyclic aromatic hydrocarbons binding to dissolved humic materials, *Environ. Sci. Technol.* **20,** 1162–6.

Gauthier, T. D., Seitz, W. R. and Grant, C. L. (1987). Effects of structural and compositional variations of dissolved humic materials on pyrene K_{oc} values, *Environ. Sci. Technol.* **21,** 243–8.

GEMS/WHO (1984). *Urban Air Pollution 1973-1980.* Global Environment Monitoring System/United Nations Environment Programme/World Health Organization, Geneva, 96 pp.

GEMS/WHO (1988). *Assessment of Urban Air Quality*. Global Environment Monitoring System/United Nations Environment Programme/World Health Organization, Monitoring and Assessment Research Centre, King's College, London, 100 pp.

Georgopoulos, P. G. and Seinfeld, J. H. (1982). Statistical distributions of air pollutant concentrations, *Environ. Sci. Technol.* **7,** 401A.

Geyer, H. J., Scheunert, I. and Korte, F. (1987). Correlation between the bio-concentration potential of organic environmental chemicals in humans and their *n*-octand/water coefficients, *Chemosphere* **16,** 239–52.

Gibbs, R. J. (Ed.) (1977). *Transport Processes in Lakes and Oceans,* Plenum Press, New York, 288 pp.

Gibson, D. T. (1977). Biodegradation of aromatic petroleum hydrocarbons. In: *Fate and effects of petroleum hydrocarbons in marine organisms and ecosystems,* D. A. Wolfe (Ed.) (Chapter 4), 36–46.

Giddings, J. M., Herbes, S. E. and Gehrs, C. W. (1985). Coal liquefaction products, *Environ. Sci. Technol.* **19,** 14–18.

GLC/SSB (1985/6). *Nitrogen Dioxide. Environmental Health Aspects of the 1984/85 London Diffusion Tube Survey.* London Environmental Supplement No. 14, Winter 1985/6, Pollution Monitoring Group, Greater London Council Scientific Services Branch (now London Scientific Services), 22 pp.

Go, F. C. (1988). *Environmental Impact Assessment, Operational Cost Benefit Analysis,* MARC Report No. 42. Monitoring and Assessment Research Centre, King's College, University of London, 60 pp.

Gobas, F. A. P. C., Opperhuizen, A. and Hutzinger, O. (1986). Bioconcentration of hydrophobic chemicals in fish: relationship with membrane permeation, *Environ. Toxicol. Chem.* **5,** 637–46.

Goerlitz, D. F., Troutman, D. E., Godsy, E. M. and Franks, B. J. (1985). Migration of wood-preserving chemicals in contaminated groundwater in a sand aquifer at Pensacola, Florida, *Environ. Sci. Technol.* **19,** 955–61.

Goldberg, L. (Ed.) (1983). Structure–activity correlation as a predictive tool in toxicology (fundamentals, methods and applications). Hemisphere Publishing Corporation and McGraw-Hill (*Chemical Industry Institute of Toxicology series*) 330 pp.

Goodall, D. W. (1972). Building and testing ecosystem models. In: *Mathematical Models in Ecology, 12th Symp. Br. Ecol. Soc.,* Grange-over-sands, Lancs, March 1971, J. N. R. Jeffers (Ed.), Blackwell, Oxford, 173–94.

Gordon, M. and Scantlebury, G. R. (1964). Non-random polycondensation: Statistical theory of the substitution effect, *Trans. Faraday Soc.* **60,** 604–21.

Gorham, E. (1989). Scientific understanding of ecosystem acidification: A historical review, *Ambio* **18,** 150–4.

Goring, C. A. I., Laskowski, D. A., Hamaker, J. W. and Meikle, R. W. (1975). Principles of pesticide degradation in soil. In: *Environmental Dynamics of Pesticides,* R. Haque and V. H. Freed (Eds), Plenum Press, New York.

Gossett, J. M. (1987). Measurement of Henry's Law Constants for C_1 and C_2 chlorinated hydrocarbons, *Environ. Sci. Technol.* **21,** 202–8.

Gray, T. R. G. (1978). Microbial aspects of the soil, plant, aquatic, air and animal environment. In: *Pesticide Microbiology,* I. N. Hill and S. J. L. Wright (Eds), Academic Press, New York.

Greenwald, H. L., Kice, E. B., Kenly, M. and Kelly, J. (1961). Determination of the distribution of non-ionic surface active agents between water and iso-octane, *Anal. Chem.* **33,** 465–8.

Grenney, W. J., Wagnet, R. J. and Jurinak, J. J. (1979). An atmospheric–

terrestrial heavy metal transport model—II. Process equations, *Ecol. Modelling* **6,** 273–88.

Guenzi, W. D. and Beard, W. E. (1974). Volatilization of pesticides. In: *Pesticides in Soil and Water,* W. D. Guenzi (Ed.), Soil. Sci. Soc. Am., Madison, WI (Chapter 6).

Guggenheim, E. A. (1967). *Thermodynamics,* North-Holland, Amsterdam.

Gunther, F. A. and Gunther, J. D. (Eds) (1983). Residues of pesticides and other contaminants in the total environment, *Residue Rev.* **85,** Springer-Verlag, 307 pp.

Gydesen, H. (1984). Mathematical models of the transport of pollutants in ecosystems, *Ecol. Bull.* (Stockholm) **36,** 17–25.

Haag, W. R. and Hoigné, J. (1986). Singlet oxygen in surface waters. 3. Photochemical formation and steady-state concentrations in various types of waters, *Environ. Sci. Technol.* **20,** 341–8.

Hales, J. M. (1978). Wet removal of sulfur compounds from the atmosphere, *Atmos. Environ.* **12,** 389–99.

Hall, L. H. and Kier, L. B. (1981). The relation of molecular connectivity to molecular volume and biological activity, *Eur. J. Med-Chem.—Chimica Therapeutica, Sept.–Oct. 1981,* **16,** 399–407.

Hall, L. H., Kier, L. B. and Murray, W. J. (1975). Molecular connectivity. II: Relationship to water solubility and boiling point, *J. Pharm. Sci.* **64,** 1974–7.

Hamaker, J. W. (1972*a*). Decomposition: Quantitative aspects. In: *Organic Chemicals in the Soil Environment,* Vol. 1, C. A. I. Goring and J. W. Hamaker (Eds), Marcel Dekker, New York.

Hamaker, J. W. (1972*b*). Diffusion and volatilization. In: *Organic Chemicals in the Soil Environment,* Vol. 1, C. A. I. Goring and J. W. Hamaker (Eds), Marcel Dekker, New York (Chapter 5).

Hamaker, J. W. (1975). In: *Environmental Dynamics of Pesticides,* R. Haque and V. H. Freed (Eds), Plenum Press, New York, 21–30.

Hammond, A. L. (1977). Oceanography: Geochemical traces offer new insight, *Science,* **195,** 164–6.

Hammond, M. W. and Alexander, M. (1972). Effects of chemical structure on microbial degradation of methyl-substituted aliphatic acid, *Environ. Sci. Technol.* **6,** 732.

Hand, V. C. and Williams, G. K. (1987). Structure–activity relationships for sorption of linear alkylbenzenesulfonates, *Environ. Sci. Technol.* **21,** 370–3.

Hansch, C. (1968). Use of substituent constants in drug modification, *Farmaco, Sci.* **23,** 294.

Hansch, C. (1969). The quantitative approach to biochemical structure–activity relationships, *Acta Chem. Res.,* **2,** 232–40.

Hansch, C. and Clayton, J. M. (1973). Lipophilic character and biological activity of drugs, II, The parabolic case, *J. Pharm. Sci.* **62,** 1–21.

Hansch, C. and Fujita, T. (1964). Rho-sigma-pi analysis. A method for the correlation of biological activity and chemicals structure, *J. Am. Chem. Soc.* **86,** 1616–26.

Hansch, C. and Leo, A. J. (1979). *Substituent Constants for Correlation Analysis in Chemistry and Biology.* John Wiley, Chichester, UK.

Hansch, C., Leo, A. and Nikaitani, D. (1972). On the additive–constitutive character of partition coefficients, *J. Org. Chem.* **37,** 3090–2.

Hansch, C., Leo, A., Unger, S., Kim, K. H., Nikaitani, D. and Lien, E. J. (1973). 'Aromatic' substituent constants for structure–activity correlations, *J. Med. Chem.* **16,** 1207–16.

Haque, R. and Freed, V. H. (1974). Behaviour of pesticides in the environment. 'Environmental Chemodynamics'. In: *Residue Rev.* **52,** 89–116, F. A. Gunther and J. D. Gunther (Eds).

Harleman, D. R. F. and Rumer, R. R. (1962). The dynamics of salt-water intrusion in porous media, *Civ. Eng. Dep. No. 55,* MIT Press, Cambridge, MA.

Harris, J. C. (1982*a*). Rate of hydrolysis. In: *Handbook of Chemical Property Estimation Methods—Environmental Behaviour of Organic Compounds,* W. J. Lyman, W. F. Rheel and D. H. Rosenblatt (Eds), McGraw-Hill (Chapter 7), 48 pp.

Harris, J. C. (1982*b*). Rate of aqueous photolysis. In: *Handbook of Chemical Property Estimation Methods—Environmental Behaviour of Organic Compounds,* W. J. Lyman, W. F. Rheel and D. H. Rosenblatt (Eds), McGraw-Hill (Chapter 8), 43 pp.

Harris, J. C. and Hayes, M. J. (1982). Acid dissociation constant. In: *Handbook of Chemical Property Estimation Methods—Environmental Behaviour of Organic Compounds,* W. J. Lyman, W. F. Rheel and D. H. Rosenblatt (Eds), McGraw-Hill (Chapter 6), 28 pp.

Harrison, R. M. (Ed.) (1983).*Pollution: Causes, Effects and Control.* Based on papers given at a residential school organized by the continuing Education Committee of the Royal Society of Chemistry, University of Lancaster, September 1982, The Royal Society of Chemistry, Special Publication No. 44, 322 pp.

Hartley, G. S. (1969). Evaporation of pesticides. In: *Pesticidal Formulations Research, Physical and Colloidal Chemical Aspects,* Advances in Chemistry Series, 86, ACS, Washington, DC (Chapter II).

Hassett, J. P. and Milicic, E. (1985). Determination of equilibrium and rate constants for binding of a polychlorinated biphenyl congener by dissolved humic substances, *Environ. Sci. Technol.* **19,** 638–43.

Hauck, R. D. and Stephenson, H. F. (1964). Fertilizer nitrogen sources—nitrification of triazine nitrogen, *J. Agric. Food Chem.* **12,** 147.

Hautala, R. R. (1978). *Surfactant effects on pesticide photochemistry in soil and water.* EPA-600/3-78-060.

Hay, A. (1988). How to identify a carcinogen, *Nature,* **332,** 782–3.

Hayduk, W. and Laudie, H. (1974). Prediction of diffusion coefficients for non-electrolysis in dilute aqueous solutions, *AIChE J.* **20,** 611–15.

Heiffer, J. L. (1980). *Air Resources Laboratories Atmospheric Transport and Dispersion Model* (*ARL-ATAD*). Air Resources Laboratories, Silver Spring, MD, ERL ARL-81.

Helling, C. S., Kearney, P. C. and Alexander, M. (1971). Behaviour of pesticides in soil, *Adv. Agron.* **23,** 147.

Helpinstill, J. G. and Van Winkle, M. (1968). *Ind. Eng. Chem. Process Des. Dev.* **7,** 213–20.

Henderson, M. E. K. (1975). The metabolism of methoxylated aromatic compound by soil fungi, *J. Gen. Microbiol.* **16,** 686–95.

Henriksen, A. and Brakke, D. F. (1988). Sulfate deposition to surface waters, *Environ. Sci. Technol.* **22**(1) 8–14.

Henze, H. R. and Blair, C. M. (1934). The number of structural isomers of the more important types of aliphatic compounds, *J. Am. Chem. Soc.* **56,** 157.

Hering, J. G. and Morel, F. M. M. (1988). Kinetics of trace metal complexation: role of alkaline-earth metals, *Environ. Sci. Technol.* **22,** 1469–78.

Hermens, J., Könemann, H., Leeuwangh, P. and Musch, A. (1985). Quantitative structure–activity relationships in aquatic toxicity studies of chemical and complex mixtures of chemicals, *Environ. Toxicol. Chem.* **4,** 273–9.

Herrick, C. J., Goodman, E. D., Guthrie, C. A., Blythe, R. H., Hendrix, G. A., Smith, R. L. and Galloway, J. E. (1982). A model of mercury contamination in a woodland stream, *Ecol. Modelling,* **15,** 1–28.

Herrick, E. C. and King, J. A. (1979). *Catalog of Organic Chemical Industries Unit Processes.* MITRE Technical Report MTR-7823, The MITRE Corp., McLean, VA, April 1979.

Herz, W. (1909). '*Der Verteilungssatz*'. Ferdinand Enke, Stuttgart, 1909, 5.

Hickman, G. T. and Novak, J. T. (1989). Relationship between subsurface biodegradation rates and microbial density, *Environ. Sci. Technol.* **23,** 525–32.

Hill, I. R. (1978). Microbial transformation of pesticides. In: *Pesticide Microbiology,* I. R. Hill and S. J. L. Wright (Eds), Academic Press, New York.

Hill, I. R. and McCarty, P. L. (1967). Anaerobic degradation of selected chlorinated pesticides, *J. Water Poll. Control. Fed.* **39,** 1259.

Hill, J., Kollig, H. P., Parris, D. F., Wolfe, N. L. and Zepp, R. G. (1976). *Dynamic behaviour of vinyl chloride in aquatic ecosystems.* EPA-600/3-76-001.

Hirsch, P. and Alexander, M. (1960). Microbial decomposition of halogenated propionic and acetic acids, *Can. J. Microbiol.* **6,** 241–9.

Hobbie, J. E., Holm-Hansen, O., Packard, T. T., Pomeroy, L. R., Sheldon, R. W., Thomas, J. P. and Wiebe, W. J. (1972). A study of the distribution and activity of micro-organisms in ocean water, *Limnol. Oceanogr.* **17,** 544.

Holm-Hansen, O. (1972). The distribution and chemical composition of particulate material in marsh and freshwaters, *Mem. Ist. Ital. Idrobiol. Dolt. Marco de Marchi,* **29** (Suppl.) 37.

Honeyman, B. D. and Santschi, P. H. (1988). Metals in aquatic systems, *Environ. Sci. Technol.* **22,** 862–71.

Hosoya, H. (1971). Topological Index. A newly proposed quantity characterizing the topological nature of structural isomers of saturated hydrocarbons, *Bull. Chem. Soc. Japan,* **44,** 2332–9.

Houghton, J. T. (1977). *The Physics of Atmospheres.* Cambridge University Press.

Huddleston, R. L. and Allred, R. C. (1963). Microbial oxidation of sulphonate alkylbenzenes, *Develop. Ind. Microbiol.* **4,** 24–38.

Hunn, J. B., Cleveland, L. and Little, E. E. (1987). Influence of pH and aluminium on developing brook trout in a low calcium water, *Environ. Pollut.* **43,** 63–73.

Hushon, J. M., Klein, A. W., Strachan, W. J. M. and Schmidt-Bleek, F. (1983). Use of OECD premarket data in environmental analysis for new chemicals, *Chemosphere* **12,** 887–910.

Hwang, H. M., Hodson, R. E. and Lee, R. F. (1986). Degradation of phenol and chlorophenols by sunlight and microbes in estuarine water, *Environ. Sci. Technol.* **20,** 1002–7.

IAEA (1980). Generic models and parameters for assessing the environmental transfer of radionuclides in predicting exposures to critical groups from routine releases. International Atomic Energy Agency, Division of Nuclear Safety and Environmental Protection, Vienna, November 1980. (Draft—not for external distribution.)

Jannasch, H. W. (1979). The ultimate sink: keynote address. In: *Proc. Workshop: Microbial degradation of pollutants in marine environments,* EPA-600/9-79/012, Washington, DC.

Jenkins, S. H. (Ed.) (1982). Water pollution research and control, Part 4, *Water Sci. Technol.* **14,** 9–11.

Johanson, R. C., Imhoff, J. C. and Davis, H. H., Jr (1980). *User's Manual for Hydrological Simulation Program—FORTRAN* (*HSPF*). EPA-600/9-80-015, US EPA Environ. Res. Lab., Athens, GA, 678 pp.

Johnsen, S. and Gribbestad, I. S. (1988). Influence of humic substances on the formation of chlorinated polycyclic aromatic hydrocarbons during chlorination of polycyclic aromatic hydrocarbon polluted water, *Environ. Sci. Technol.* **22,** 978–81.

Johnson, H. E. and Parrish, R. (Chairmen) (1978). Discussion session synopsis. In: *Estimating the Hazard of Chemical Substances to Aquatic Life,* J. Cairns, Jr, K. L. Dickson and A. W. Maki (Eds), ASTM STP 657, 71–7.

Johnson, W. B., Ludwig, F. L., Dabberdt, W. F. and Allen, R. J. (1973). An urban diffusion simulation model for carbon monoxide, *J. Air. Pollut. Control Assoc.* **23,** 490–8.

Jörgensen, S. E. (Ed.) (1979). Modelling the distribution and effect of toxic substances, Special issue, *Ecol. Modelling,* **6**(3), 179–288.

Jörgensen, S. E. (Ed.) (1983). *Application of Ecological Modelling in Environmental Management, Part A: Developments in Environmental Modelling,* Vol. 4A, Elsevier, Amsterdam/Oxford/New York, 735 pp.

Jörgensen, S. E. (Ed.) (1984). *Modelling the Fate and Effect of Toxic Substances in the Environment,* Proc. Symp., June 1983, Developments in Environmental Monitoring, Vol. 6, Copenhagen, Denmark, Elsevier Science Publications, B. V. Amsterdam, 342 pp. (Also as *Ecol. Modelling* **22,** 1983/84.)

Jörgensen, S. E. and Johnsen, I. (1981). Principles of environmental science and technology, *Stud. Environ. Sci.* **14,** Elsevier, 516 pp.

Josephson, J. (1983). Subsurface organic contaminants, *Environ. Sci. Technol.* **17,** 518A–21A.

Josephson, J. (1985). Implementing Superfund, *Environ. Sci. Technol.* **20,** 23–8.

JRB (1980). *Methodology for rating the hazard potential of waste disposal sites.* JRB Associates, McLean, VA.

Junge, C. E. (1974). Residence time and variability of tropospheric trace gases, *Tellus* **26,** 477–88.

Junge, C. E. (1977). Basic considerations about trace constituents in the atmosphere as related to the fate of global pollutants. in: *Part 1: Fate of Pollutants in the Air and Water Environments,* I. H. Suffet (Ed.), John Wiley and Sons, New York, pp. 7–25.

Jurs, P. C. (1983). Studies of relationships between molecular structure and biological activity by pattern recognition methods. In: *Structure–Activity Correlation as a Predictive Tool in Toxicology,* L. Goldberg (Ed.), Hemisphere Publishing Corporation and McGraw-Hill, 93–110.

Jury, W. A. (1982). Simulation of solute transport using a transfer function model, *Water Res.* **18,** 363–8.

Jury, W. A., Grover, R., Spencer, W. F. and Farmer, W. J. (1980). Modeling vapour losses of soil-incorporated Triallate, *Soil Sci. Soc. Am. J.* **44,** 445–50.

Jury, W. A., Spencer, W. F. and Farmer, J. W. (1983). Behaviour assessment model for trace organics in soil. 1. Model description, *J. Environ. Qual.* **12,** 558–64.

Kalmaz, E. V. and Kalmaz, G. D. (1979). Transport, distribution and toxic effects of polychlorinated biphenyls in ecosystems: Review, *Ecol. Modelling* **6,** 223–51. In: S. E. Jorgensen (Ed.) (1979).

Karickhoff, S. W., Brown, D. S. and Scott, T. A. (1979). Sorption of hydrophobic pollutants in natural sediments, *Water Res.* **13,** 241–8.

Kaufman, D. D. (1966). *Structure of pesticides and decomposition by micro-organisms. Pesticides and their effects in soil and water.* ASA Special Publication No. 8, Soil. Sci. Soc. Amer., Madison, WI. (Cited in Alexander (1973).)

Kaufman, D. D. and Hornick, S. B. (1984). Movement and fate of hazardous materials in soil systems. In: *Proc. Conf. on Risk and Decision Analysis for Hazardous Waste Disposal,* Hazardous Materials Control Research Institute, Silver Spring, MD.

Kearney, P. C. and Plimmer, J. R. (1970). Relation of structure to pesticide decomposition. In: *Pesticides in the Soil,* Michigan State University, E. Lansing, MI.

Keddie, A. W. C. (1983). Atmospheric dispersal of pollutants and the modelling of air pollution. In: *Pollution: Causes, Effects and Control,* R. M. Harrison (Ed.), Royal Soc. Chem. Special Publ. No. 44 (Chapter 10), 191–207.

Kenaga, E. E. (1968). *Down to Earth,* **23,** 11–18.

Kenaga, E. E. and Goring, C. A. I. (1980). Relationship between water solubility, soil sorption, octanol–water partitioning and bioconcentration of chemicals in biota. In: *Aquatic Toxicology,* J. R. Eaton, P. R. Parish and A. Hendricks (Eds), ASTM, Philadelphia, 78–115.

Kerfoot, H. B. and Barrows, L. J. (1986). *Soil Gas Measurement for Detection of Subsurface Organic Contamination,* US Environmental Protection Agency, Las Vegas, NV, p. 2.

Kier, L. B. and Hall, L. H. (1976). *Molecular Connectivity in Chemistry and Drug Research*. Academic Press, New York/San Francisco/London, 257 pp.

Kier, L. B. and Hall, L. H. (1981). Derivation and significance of valence molecular connectivity, *J. Pharm. Sci.* **70,** 583–9.

Kier, L. B., Hall, L. H., Murray, W. J. and Randic, M. (1975). Molecular connectivity. I: Relationship to non-specific local anesthesia, *J. Pharm. Sci.* **64,** 1971–4.

Kier, L. B., Hall, L. H., Murray, W. J. and Randic, M. (1976). Molecular Connectivity. V: Connectivity series concept applied to density, *J. Pharm. Sci.* **65,** 1226–30.

Kinniburgh, D. G. (1986). General purpose adsorption isotherms, *Environ. Sci. Technol.* **20,** 895–904.

Kinzelbach, W. (1986). *Groundwater Modelling: An Introduction with Sample Programs in BASIC*. Elsevier Science Publications, Amsterdam. (Errata available from IGWMC, Indianapolis Office.)

Kirda, C., Nielsen, D. R. and Biggar, J. W. (1973). Simultaneous transport of chlorides and water during infiltration, *Soil Sci. Soc. Am. Proc.* **37,** 339–45.

Kleopfer, R. D., Easley, D. M., Hass, B. B. Jr, Deihl, J. G., Jackson, D. E. and Wurrey, C. J. (1985). Anaerobic degradation of trichloroethylene in soil, *Environ. Sci. Technol.* **19,** 277–80.

Klopffer, W., Rippen, G. and Frische, R. (1982). Physiochemical properties as useful tools for predicting the environmental fate of organic chemicals, *Ecotoxicol. Environ. Safety* **6,** 294–301.

Koch, R. (1982*a*). Molecular connectivity and acute toxicity of environmental pollutants, *Chemosphere* **11,** 925–31.

Koch, R. (1982*b*). Molecular connectivity index for assessing ecotoxicological behaviour of organic compounds, *Toxicol. Environ. Chem.* **6,** 87–96.

Koeman, J. H. (1982). Ecotoxicological evaluation: the eco-side of the problem, *Ecotoxicol. Environ. Safety* **6,** 358–62.

Könemann, H. and van Leeuwen, K. (1980). Toxico-kinetics in fish—accumulation and elimination of 6 chlorobenzenes by guppies, *Chemosphere* **9,** 3–19.

Konikow, L. F. and Bredehoeft, J. D. (1988). *Computer Model of Two-dimensional Solute Transport and Dispersion in Groundwater* (*MOC*). US Geological Survey.

Kormann, C., Bahnemann, D. W. and Hoffman, M. R. (1988). Photocatalytic production of H_2O_2 and organic peroxides in aqueous suspensions of TiO_2, ZnO and desert sand, *Environ. Sci. Technol.* **22,** 798–806.

Kramer, N. and Doetsch, R. N. (1950). Growth of phenol-utilizing bacteria on aromatic carbon sources, *Arch. Biochem. Biophys.* **26,** 401–5.

Kringstad, K. P. and Lindström, K. (1984). Spent liquors from pulp bleaching, *Environ. Sci. Technol.* **18,** 236A–48A.

Ladd, J. N. (1956). The oxidation of hydrocarbons by soil bacteria: I. Morphological and biochemical properties of a soil diphtheroid utilizing hydrocarbons, *Aust. J. Biol. Sci.* **9,** 92–104.

Lam, D. C. L., Minns, C. K., Hodson, P. V., Simons, T. J. and Wong, P. (1976). Computer model for toxicant spills in Lake Ontario. In: *Environmental Biogeochemistry 2. Metals Transfer and Ecological Mass Balance,* J. O. Nriagu (Ed.) (Chapter 46), 743–61.

Larson, R. A. and Berenbaum, M. R. (1988). Environmental phototoxicity, *Environ. Sci. Technol.* **22,** 354–60.

Lassetre, E. N. (1937). The hydrogen bond and association, *Chem. Rev.* **20,** 259–303.

Lassiter, R. R. (1982). In: *Modelling the Fate of Chemicals in the Aquatic Environment.* Ann Arbor Science Publications, Ann Arbor, MI, 287–301.

Lassiter, R. R., Baughman, G. L. and Burns, L. A. (1979). *Fate of Toxic Organic Substances in the Aquatic Environment.* US EPA, Environ. Res. Lab., Athens, GA.

Laughlin, R. B. Jr., Johannesen, R. B., French, W., Guard, H. and Brinckman, F. E. (1985). Structure–activity relationships for organotin compounds, *Environ. Toxicol. Chem.* **4,** 342–51.

Lavelle, J. W., Ozturgut, E., Baker, E. T., Tennant, D. A. and Walker, S. L. (1988). Settling speeds of sewage sludge in seawater, *Environ. Sci. Technol.* **22,** 1201–7.

Lawrence, G. B. and Driscoll, C. T. (1988). Aluminium chemistry downstream of a whole-tree-harvested watershed, *Environ. Sci. Technol.* **22,** 1293–9.

Laxen, D. P. H. (1983*a*). Cadmium adsorption in freshwaters—a quantitative appraisal of the literature, *Sci. Tot. Environ.* **30,** 129–46.

Laxen, D. P. H. (1983*b*). The chemistry of metal pollutants in water. In: *Pollution: Causes, Effects and Control,* R. M. Harrison (Ed.), Roy. Soc. Chem., Special Publ. No. 44 (Chapter 6), 104–23.

Laxen, D. P. H. and Thompson, M. A. (1987). Sulphur dioxide in Greater London, 1931–1985, *Environ. Pollut.* **43,** 103–14.

Leadbetter, M. R. and Tucker, W. G. (1981). Environmental assessment of industrial discharges based on multiplicative models, *Environ. Sci. Technol.* **15,** 1355–60.

LeGrand, H. E. (1980). *A standardized system for evaluating waste disposal sites.* Natl. Water Well Assoc., Washington, DC.

Leo, A. (1975). *Symposium on structure–activity correlations in studies of toxicity and bioconcentration with aquatic organisms.* Great Lakes Research Advisory Board, Burlington, Ontario, p. 151.

Leo, A. and Hansch, C. (1971). Linear free-energy relationships between partitioning solvent systems, *J. Org. Chem.* **36,** 1539–44.

Leo, A. J. and Hansch, C. (1986). *CLOGP3 Users Guide MedChem Software Release 3.32.* Pomona College, Claremont, CA.

Leo, A., Hansch, C. and Elkins, D. (1971). Partition coefficients and their use, *Chem. Rev.* **71,** 525–616.

Leone, J. A. and Seinfeld, J. H. (1984). Analysis of the characteristics of complex chemical reaction mechanisms: Application to photochemical smog chemistry, *Environ. Sci. Technol.* **18,** 280–7.

Lepkowski, W. (1985). Bhopal—India city begins to heal but conflicts remain, *Chem. Eng. News* **63,** 18–32.

Lerman, A. (1978). Chemical exchange across sediment water interface, *Ann. Rev. Earth Planet. Sci.* **6,** 281–303.

Leuenberger, C., Ligocki, M. P. and Pankow, J. F. (1985). Trace organic compounds in rain. 4. Identities, concentrations, and scavenging mechanisms for phenols in urban air and rain, *Environ. Sci. Technol.* **19,** 1053–8.

Lewis, D. L., Kollig, H. P. and Hodson, R. E. (1986). Nutrient limitation and adaptation of microbial populations to chemical transformations, *Appl. Environ. Microbiol.* **51,** 598–603.

Lewis, G. N. (1901*a*). *Proc. Am. Acad.* **37,** 49.

Lewis, G. N. (1901*b*). *Z. Physik. Chem.* **38,** 205.

Lien, E. J. (1983). Letter to the editor, *Chem. Eng. News.* **61,** 2.

Lien, E. J. (1985). Molecular structures and different modes of exposure affecting transport and toxicities of chemicals: QSAR analysis, *Environ. Toxicol. Chem.* **4,** 259–71.

Lindström, K., Nordin, J. and Österberg, R. (1981). In: *Advances in the Identification and Analysis of Organic Pollutants in Water,* L. H. Keith (Ed.), Ann Arbor Science Publications, Ann Arbor, MI.

Lipnick, R. L., Johnson, D. E., Gilford, J. H., Bickings, C. K. and Newsome, L. D. (1985). Comparison of fish toxicity screening data for 55 alcohols with the quantitative structure–activity relationship prediction of minimum toxicity for non reactive non electrolyte organic compounds, *Environ. Toxicol. Chem.* **4,** 281–96.

Liss, P. S. and Slater, P. G. (1974). Flux of gases across the air–sea interface, *Nature* (*London*) **247,** 181–4.

Liss, P. S. and Slinn, W. G. N. (1983). *Air–Sea exchange of gases and particles.* NATO-ASI Series 108, Reidel, Boston, MA.

Lioy, P. J. and Daisey, J. M. (1986). Airborne toxic elements and organic substances, *Environ. Sci. Technol.* **20,** 8–14.

Long, M. I. E. and Winsor, G. W. (1960). Isolation of some urea–formaldehyde compounds and their decomposition in soil, *J. Sci. Food Agric.* **11,** 441.

Lowenthal, D. H. and Rahn, K. (1988). Tests of regional elemental tracers of pollution aerosols. 2. Sensitivity of signatures and apportionments to variations in operating parameters, *Environ. Sci. Technol.* **22,** 420–6.

Lowenthal, D. H., Wunschel, K. R. and Rahn, K. A. (1988). Tests of regional elemental tracers of pollution aerosols. 1. Distinctions of regional signatures, stability during transport, and empirical validation, *Environ. Sci. Technol.* **22,** 413–20.

Lu, P. Y. and Metcalf, R. L. (1975). Environmental fate and biodegradability of benzene-derivatives as studies in a model aquatic ecosystem, *Environ. Health Perspect.* **10,** 269–84.

Ludlow, L. and Smit, B. (1987). Assessing the implications of environmental change for agricultural production—The case of acid rain in Ontario, Canada, *J. Environ. Manage.* **25,** 27–44.

Lurmann, F. W., Godden, D. A. and Collins, H. M. (1985). *Users Guide to the PLMSTAR Air Quality Simulation Model.* ERT Document M-2206–100, Environmental Research and Technology, Newbury Park, CA.

Lvovitch, M. I. (1977). *Ambio* **6,** 13.

Lyman, W. J. (1982*a*). Solubility in water. In: *Handbook of Chemical Property Estimation Methods—Environmental Behaviour of Organic Compounds,* W. J. Lyman, W. F. Reehl and D. H. Rosenblatt (Eds), McGraw-Hill (Chapter 2), 52 pp.

Lyman, W. J. (1982*b*). Atmospheric residence time. In: *Handbook of*

Chemical Property Estimation Methods—Environmental Behaviour of Organic Compounds, W. J. Lyman, W. F. Reehl and D. H. Rosenblatt (Eds), McGraw-Hill (Chapter 10), 33 pp.

Mackay, D. (1979). Finding fugacity feasible, *Environ. Sci. Technol.* **13,** 1218–23.

Mackay, D. (1982). Correlation of bioconcentration factors, *Environ. Sci. Technol.* **16,** 274–8.

Mackay, D. and Hughes, A. I. (1984). Three-parameter equation describing the uptake of organic compounds by fish, *Environ. Sci. Technol.* **18,** 439–44.

Mackay, D. and Paterson, S. (1981). Calculating fugacity, *Environ. Sci. Technol.* **15,** 1006–14.

Mackay, D. and Paterson, S. (1984). *A fugacity model of atmosphere pathways of organic contaminants to the Great Lakes,* Report prepared for Environment Canada under DSS Contract No. OSE83-00176, 63 pp (inc. 20 pp comprising tables, bibliography and annex).

Mackay, D. and Shiu, W. (1975). *Chemistry and physics of aqueous gas solutions*. Electrochemical Society, New York, p. 93.

Mackay, D. and Wolkoff, A. W. (1973). Rate of evaporation of low solubility contaminants from water bodies to atmosphere, *Environ. Sci. Technol.* **7,** 611–14.

Mackay, D. and Yuen, T. K. (1979). Volatilization rates of organic contaminants from rivers. *Proc. 14th Can. Symp. 1979,* Water Pollution Research, Canada.

Mackay, D., Bobra, A., Shiu, W. Y. and Yalkowsky, S. H. (1980*a*). Relationship between aqueous solubility and octanol–water partition-coefficients, *Chemosphere* **9,** 701.

Mackay, D., Mascavenhas, R., Shiu, W. Y., Valvani, S. C. and Yalkowsky, S. H. (1980*b*). Aqueous solubility of polychlorinated-biphenyls, *Chemosphere* **9,** 257–64.

Mackay, D., Paterson, S. and Joy, M. (1983*a*). Application of fugacity models to the estimation of chemical distribution and persistence in the environment. In: *Fate of Chemicals in the Environment,* R. L. Swann and A. Eschenroeder (Eds), ACS. Symp. Ser. 225 (Chapter 9), 175–96.

Mackay, D., Joy, M. and Paterson, S. (1983*b*). A quantitative water, air, sediment interaction (QWASI) fugacity model for describing the fate of chemicals in lakes, *Chemosphere* **12,** 981–97.

Mackay, D., Paterson, S. and Joy, M. (1983*c*). A quantitative water, air, sediment interaction (QWASI) fugacity model for describing the fate of chemicals in rivers, *Chemosphere* **12,** 1193–208.

Mackay, D., Shiu, W. Y., Billington, J. and Huang, G. L. (1983*d*). In: *Physical Behaviour of PCB's in the Great Lakes,* D. Mackay, S. Paterson, S. J. Eiserleich and M. S. Simmons (Eds), Ann Arbor Science Publications, Ann Arbor, MI.

Mackay, D. M., Roberts, O. V. and Cherry, J. A. (1985). Transport of organic contaminants in groundwater, *Environ. Sci. Technol.* **19,** 384–92.

Mackay, D., Paterson, S. and Schroeder, W. H. (1986). Model describing the rates of transfer processes of organic chemicals between atmosphere and water, *Environ. Sci. Technol.* **20,** 810–16.

Macrae, I. C. and Alexander, M. (1963). Metabolism of phenoxyalkyl carboxylic acids by a Flavobacterium species, *J. Bacteriol.* **86,** 1231.

Mailhot, H. (1987). Prediction of algal bioaccumulation and uptake rate of nine organic compounds by ten physicochemical properties, *Environ. Sci. Technol.* **21,** 1009–13.

Mailhot, H. and Peters, R. H. (1988). Empirical relationships between the 1-octanol/water partition coefficient and nine physicochemical properties, *Environ. Sci. Technol.* **22,** 1479–88.

Malcolm, R. L. and MacCarthy, P. (1986). Limitations in the use of commercial humic acids in water and soil research, *Environ. Sci. Technol.* **20,** 904–11.

Mamantov, A. (1984). Linear correlation between photolysis rates and toxicity of polychlorinated dibenzo-*p*-dioxins, *Environ. Sci. Technol.* **18,** 808–10.

Manahan, S. E. (1969). *Environmental Chemistry*. Willard Grant Press, Inc., Boston.

Manheim, F. J. (1970). The diffusion of ions in unconsolidated sediments, *Earth Planet Sci. Lett.* **9,** 307–9.

Mantoura, R. F. C., Dickson, A. and Riley, J. P. (1978). Complexation of metals with humic materials in natural waters, *Estuar. Coast. Mar. Sci.* **6,** 387–408.

Marinsky, J. A. and Ephraim, J. (1986). A unified physicochemical description of the protonation and metal ion complexation equilibria of natural organic acids (humic and fulvic acids). 1. Analysis of the influence of polyelectrolyte properties on protonation equilibria in ionic media: Fundamental concepts, *Environ. Sci. Technol.* **20,** 349–54.

Marple, L., Berridge, B. and Throop, L. (1986). Measurement of the water–octanol partition coefficient of 2,3,7,8-tetrachlorodibenzo-*para*-dioxin, *Environ. Sci. Technol.* **20,** 397–9.

Marrin, D. L. (1988). *Ground Water Monit. Rev.* **8,** 51–4.

Marrin, D. L. and Kerfoot, H. B. (1988). Soil-gas surveying techniques, *Environ. Sci. Technol.* **22,** 740–5.

Martinez, M. J. (1985). *FEMTRAN—A Users Guide: SAND84-0747.* Sandia National Laboratories, Alberquerque, NM.

Maul, P. R. (1982*a*). A time-dependent model for the atmospheric transport of sulfur compound pollutants—I. Model details, *Environ. Pollut.* **B3,** 159–79.

Maul, P. R. (1982*b*). A time-dependent model for the atmospheric transport of sulfur compound pollutants—II. Applications to the long-range transport of sulfur compounds, *Environ. Pollut.* **B4,** 1–25.

Mayer, R., Letey, J. and Farmer, W. J. (1974). Models for predicting volatilization of soil-incorporated pesticides, *Soil Sci. Soc. Am. Proc.* **38,** 563–8.

Mazurek, M. A. and Simoneit, B. R. (1986). Organic components in bulk and wet-only precipitation. *CRC Crit. Rev. Env.* **16**(1), 1–140.

McAuliffe, C. D. (1977). Evaporation and solution of CO_2 to C_{10} hydrocarbons from crude oils on the sea surface. In: *Fate and effects of petroleum hydrocarbons in marine organisms and ecosystems,* D. A. Wolfe (Ed.) (Chapter 37), 363–71.

McCall, J. M. (1975). Liquid–liquid partition-coefficients by high pressure liquid chromatography, *J. Med. Chem.* **18,** 549–52.

McCall, P. J., Swann, R. L., Laskowski, D. A., Vrona, S. A., Unger, S. M. and Dishburger, H. J. (1981). Prediction of chemical mobility in soil from sorption coefficients. In: *Aquatic Toxicology and Hazard Assessment: 4th Conf.,* D. R. Branson and K. L. Dickson (Eds), Am. Soc. for Testing and Materials, ASTM STP 737, 49–58.

McCall, P. J., Laskowski, D. A., Swann, R. L. and Dishburger, N. J. (1983*a*). Estimation of environmental partitioning of organic chemicals in model ecosystems. In: *Residue Rev.* **85,** F. A. Gunter and J. D. Gunter (Eds), 231–44.

McCall, P. J., Swann, R. L. and Laskowski, D. A. (1983*b*). Partition models for equilibrium distribution of chemicals in environmental compartments. In: *Fate of Chemicals in the Environment–Compartmental and Multimedia Models for Predictions,* R. L. Swann and A. Eschenroeder (Eds), ACS Symp. Ser., 225 (Chapter 6), 105–23.

McCarthy, J. F. and Jimenez, B. D. (1985). Interactions between polycyclic aromatic hydrocarbons and dissolved humic material: Binding and dissociation, *Environ. Sci. Technol.* **19,** 1072–6.

McConnell, E. E., Bucher, J. R., Schwetz, B. A., Gupta, B. N., Shelby, M. D., Luster, M. F., Brody, A. R., Boorman, G. A., Richter, C., Stevens, M. A. and Adkins, B., Jr (1987). Toxicity of methyl isocynate, *Environ. Sci. Technol.* **21,** 188–93.

McDonald, M. G. and Harbaugh, A. W. (1988). *A Modular Three-dimensional Finite-difference Ground-water Flow Model* (*MODFLOW*), US Geological Survey. Latest documentation in USGS Techniques of Water-Resources Investigations, Book 6, Chapter A-1. First version was Open-File Report 83-875.

McDuffie, B. (1981). Estimation of octanol–water partition-coefficients for organic pollutants using reverse-phase HPLC, *Chemosphere* **10,** 73–83.

McGowan, J. C. and Mellors, A. (1986). Molecular volumes and the toxicities of chemicals to fish, *Bull. Environ. Contam. Toxicol.* **36,** 881–7.

McIlroy, L. M., DePinto, J. V., Young, T. C. and Martin, S. C. (1986). Partitioning of heavy metals to suspended solids of the Flint River, Michigan, *Environ. Toxicol. Chem.* **5,** 609–23.

McKenna, E. J. and Heath, R. D. (1976). *Biodegradation of polynuclear aromatic hydrocarbon pollutants by soil and water micro-organisms.* Prepared for US Dept. of the Interior Water Resources Research Act of 1964.

McMahon, T. A., Denison, D. J. and Fleming, R. (1976). A long distance transportation model incorporating washout and dry deposition components, *Atmos. Environ.* **10,** 751–60.

McNaughton, D.·J. (1980). Initial comparison of SURE/MAP3S sulfur oxide observations with long-term regional model predictions, *Atmos. Environ.* **14,** 55–63.

McNaughton, D. J. and Scott, B. C. (1980). Modelling evidence of incloud transformation of sulfur dioxide in sulfate, *J. Air Pollut. Control Assoc.* **30,** 272–273.

Mercer, J. W. and Faust, (1981). *Ground water modelling*. Natl. Water Well Assoc., Washington, DC.
Metcalf, R. L. and Lu, P. Y. (1973). *Environmental distribution and metabolic fate of key industrial pollutants and pesticides in a model ecosystem*. Univ. Illinois at Urbana Water Resources Center, U1LU-WRC-0069, July 1973.
Metcalf, R. L., Sanborn, J. R., Lu, P. Y. and Nye, D. (1975). *Arch. Environ. Contam. Toxicol.* **3,** 151–65.
Meyer, H. (1989). *Arch. Exp. Pathol. Pharmakol.* **42,** 110.
MIAS (1983). Sand transport probe, *MIAS News Bill* No. **6,** 12.
Mill, T. (1981). Minimum date needed to estimate environmental fate and effects for hazard classification of synthetic chemicals. *Proc. Workshop on the control of existing chemicals,* under the patronage of the OECD, 10–12 June 1981, W. Berlin, Reichstagsgebaude, 207–27.
Mill, T. (1989). Structure–activity relationships for photooxidation processes in the environment, *Environ. Toxicol. Chem.* **8,** 31–43.
Mill, T., Hendry, D. G. and Richardson, H. (1980). Free-radical oxidants in natural waters, *Science* **207,** 886–7.
Mill, T., Mabey, W. R., Lan, B. Y. and Baraze, A. (1981). Photolysis of polycyclic aromatic-hydrocarbons in water, *Chemosphere* **10,** 1281–90.
Miller, D. R., Butler, G. and Bramall, L. (1976). Validation of ecological system models, *J. Environ. Manage.* **4,** 383–401.
Miller, G. C. and Zepp, R. G. (1979). Effects of suspended sediments on photolysis rates of dissolved pollutants, *Environ. Sci. Technol.* **13,** 860–3.
Miller, M. M., Wasik, S. P., Huang, G. L., Shiu, W. Y. and Mackay, D. (1985). Relationships between octanol–water partition coefficient and aqueous solubility, *Environ. Sci. Technol.* **19,** 522–9.
Miller, R. M., Singer, G. M., Rosen, J. D. and Bartha, R. (1988). Sequential degradation of chlorophenols by photolytic and microbial treatment, *Environ. Sci. Technol.* **22,** 1215–19.
MITRE (1981). *Site ranking model for determining remedial action priorities among the uncontrolled hazardous substances facilities*. The MITRE Co., McLean, VA.
Molina, M. J. and Rowland, F. S. (1974). Stratospheric sink for chlorofluoromethanes: chlorine atom catalysed destruction of ozone, *Nature* (*London*), **249,** 810–12.
Morel, F. M. M., Dzombak, D. A. and Fish, W. (1987) (Reply to Ruzic, I. (1987)), *Environ. Sci. Technol.* **21,** 1135–6.
Mullins, L. J. (1954). Some physical mechanisms in narcosis, *Chem. Rev.* **54,** 289–323.
Munn, R. E. and Bolin, B. (1971). Review paper. Global air pollution—meteorological aspects. A survey, *Atmos. Environ.* **5,** 363–402.
Murali, V. and Aylmore, L. A. G. (1983). Competitive adsorption during solute transport in soils: 3. A review of experimental adsorption and an evaluation of simple competition models, *Soil Sci.* **136,** 279–90.
Murphy, T. J., Pokojowczyk, J. C. and Mullin, M. D. (1983). In: *Physical Behaviour of PCBs in the Great Lakes,* D. Mackay *et al.* (Eds), Ann Arbor Science Publications, Ann Arbor, MI, 49–58.
Murphy, T. J., Formanski, L. J., Brownawell, B. and Meyer, J. A. (1985). Polychlorinated biphenyl emissions to the atmosphere in the Great Lakes

region. Municipal landfills and incinerators, *Environ. Sci. Technol.* **19,** 942–6.

Murray, W. J., Hall, L. H. and Kier, L. B. (1975). Molecular connectivity. III: Relationship to partition coefficients, *J. Pharm. Sci.* **64,** 1978–81.

Narasimhan, T. N. (1975). A unified numerical model for saturated–unsaturated groundwater flow. Thesis, University of California, Berkeley, CA.

NAS (1984). *Causes and Effects of Changes in Stratospheric Ozone.* National Academy of Sciences, Washington, DC, xi + 254 pp.

NASA (1988). *Present State of Knowledge of the Upper Atmosphere 1988: An Assessment Report.* R. T. Watson and Trends Panel, M. J. Prather and Ad Hoc Theory Panel and M. M. Kurylo and NASA panel for Data Evaluation, National Aeronautical and Space Administration, NASA Publication No. 1208.

NCAR (1985). *The NCAR Regional Acid Deposition Model.* National Center for Atmosphere Research, Boulder, CO, NCAR/TN-256 + STR.

NCAR (1986). *Preliminary Evaluation Studies with the Regional Acid Deposition Model (RADM).* National Centre for Atmospheric Research, Boulder, CO, NCAR/TN-256 + STR.

Neely, W. B. (1979). A preliminary assessment of the environmental exposure to be expected from the addition of a chemical to a simulated aquatic ecosystem, *Int. J. Environ. Studies* **13,** 101–8.

Neely, W. B. (1980). *Chemicals in the Environment, Distribution, Transport, Fate, Analysis,* Marcel Dekker, New York, 245 pp.

Neely, W. B. (1981). Complex problems—simple solutions, *Chem. Tech.* **11,** 249–51.

Neely, W. B. (1982). Review: Organising data for environmental studies, *Environ. Toxicol. Chem.* **1,** 259–66.

Neely, W. B. and Blau, G. E. (1977). In: *Pesticides in Aquatic Environments,* H. A. Q. Khan and J. P. Bederka (Eds), Plenum Press, New York, 145–63.

Neely, W. B., Branson, D. R. and Blau, G. E. (1974). Partition coefficient to measure bioconcentration potential of organic chemicals in fish, *Environ. Sci. Technol.* **8,** 1113–15.

Nelson, R. W. and Schur, J. A. (1980). *Assessment of effectiveness of geologic oscillation systems: PATHS ground water hydrologic model.* Battelle, Pacific Northwest Lab., Richland, WA.

Nernst, W. (1891). *Z. Phys. Chem.* **8,** 110.

Nesbitt, H. J. and Watson, J. R. (1980). Degradation of the herbicide 2,4-D in river water—II. The role of suspended sediment, nutrients and water temperature, *Water Res.* **14,** 1689–94.

Newell, A. D. and Sanders, J. G. (1986). Relative copper binding capacities of dissolved organic compounds in a coastal-plain estuary, *Environ. Sci. Technol.* **20,** 817–21.

Nicholson, B. C., Maguire, B. P. and Bursill, D. B. (1984). Henry's Law Constants for the trihalomethanes: Effects of water composition and temperature, *Environ. Sci. Technol.* **18,** 518–21.

Nihoul, J. C. J. (Ed.) (1978). *Hydrodynamics of Estuaries and Fjords.* Elsevier, Amsterdam, 546 pp.

Nirmalakhandan, N. and Speece, R. E. (1988). Structure–activity relationships, *Environ. Sci. Technol.* **22,** 606–15.

Nkedi-Kizza, P., Rao, P. S. C. and Hornsby, A. G. (1985). Influence of organic cosolvents on sorption of hydrophobic organic chemicals by soils, *Environ. Sci. Technol.* **19,** 975–9.

Nkedi-Kizza, P., Rao, P. S. C. and Hornsby, A. G. (1987). Influence of organic cosolvents on leaching of hydrophobic organic chemicals through soils, *Environ. Sci. Technol.* **21,** 1107–11.

NOAA (1975). *A regional-continental scale transport, diffusion, and deposition model. Part I: Trajectory Model. Part II: diffusion–deposition Models.* National Oceanic and Atmospheric Administration Technical Memorandum, ERL ARL-50, 25 pp.

Norrington, F. E., Hyde, R. M., Williams, S. G. and Wooten, R. (1975). Physico-chemical-activity relations in practice. 1: A rational and self-consistent data bank, *J. Med. Chem.* **18,** 604–7.

NRC (1984). *National Research Council Report ISBN 0-0309-03433-7.* National Research Council, Washington, DC, p. 76.

NRCC (1981). *A screen for the relative persistence of lipophilic organic chemicals in aquatic ecosystems—an analysis of the role of a simple computer model in screening.* National Res. Council Canada, No. 18570, 300 pp.

NTIS (*a*). *User's Manual for the APRAC-1A Urban Diffusion Model Computer Program.* NTIS PB 213-091.

NTIS (*b*). *A practical multipurpose urban diffusion model for carbon monoxide.* NTIS PB 196-003.

NTIS (*c*) *Field study for initial evaluation of an urban diffusion model for carbon monoxide.* NTIS PB 203–469.

NTIS (*d*) *Evaluation of the APRAC-1A Urban Diffusion Model for Carbon Monoxide.* NTIS PB 210-813.

NTIS (*e*) *User's manual for single-source (CRSTER) model.* Monitoring and Data Analysis Division US EPA, Research Triangle Park, NC. EPA-450/2-77-013, 1977, NTIS PB 271-360.

Odum, E. P. (1971). *Fundamentals of Ecology.* 3rd edn, W. B. Saunders Co., Philadelphia, PA.

Ogram, A., Sayler, G. S., Gustin, D. and Lewis, R. J. (1988). DNA adsorption to soils and sediments, *Environ. Sci. Technol.* **22,** 982–4.

OHS (1976–80). *Chemicals in the Environment.* Office of Health Studies, Japanese Environment Agency, Report series 2-6.

Oliver, B. G. (1985). Desorption of chlorinated hydrocarbons from spiked and anthropogenically contaminated sediments, *Chemosphere* **14,** 1087–106.

Oliver, B. G. and Niimi, A. J. (1983). Bioconcentration of chlorobenzenes from water by rainbow trout—correlations with partition-coefficients and environmental residues, *Environ. Sci. Technol.* **17,** 287–91.

Ollis, D. F. (1985). Contaminant degradation in water, *Environ. Sci. Technol.* **19,** 480–4.

Onishi, Y. and Wise, S. E. (1982). *Mathematical model, SERATRA, for sediment–contaminant transport in rivers and its application to pesticide transport in Four Mile and Wolf Creeks in Iowa.* EPA-600/3-82-045, Athens, GA, 56 pp.

Onishi, Y., Brown, S. M., Olsen, A. R., Parkhurst, M. A., Wise, S. E. and Walters, W. A. (1982). *Methodology for overland and instream migration and risk assessment of pesticides*. EPA-600/3-82-024, Athens, GA, 115 pp.

Opperhuizen, A., Gobas, F. A. P. C., van der Steen, J. M. B. and Hutzinger, O. (1988). Aqueous solubility of polychlorinated biphenyls related to molecular structure, *Environ. Sci. Technol.* **22,** 638–46.

Ostrenga, J. A. (1969). Correlation of biological activity with chemical structure, use of molar attraction constants, *J. Med. Chem.* **12,** 349–52.

Othmer, D. F. and Thakar, M. S. (1953). Correlating diffusion coefficients in liquids, *Ind. Eng. Chem.* **45,** 589.

Ott, W. R. (1978). *Environmental Indices—Theory and Practice*. Ann Arbor Science Publications, Ann Arbor, MI, 371 pp.

Ott, W. R. (1985). Total human exposure, *Environ. Sci. Technol.* **19,** 880–886.

Ott, W. R. and Mage, D. T. (1976). A general purpose univariate probability model for environmental data analysis, *Comput. & Ops. Res.* **3,** 209–16.

Overton, E. (1901). '*Studien uber die Narkose*'. Fisher, Jena, Germany.

Pankow, J. F., Isabelle, L. M. and Asher, W. E. (1984). Trace organic-compounds in rain. 1. Sampler design and analysis by adsorption thermal-desorption (ATD), *Environ. Sci. Technol.* **18,** 310–18.

Painter, H. A. (1974). Biodegradability, *Proc. R. Soc. Lond.* (*Ser. B*) **185,** 149–58.

Park, R. A., Connolly, C. I., Albanese, J. R., Clesceri, L. S., Heitzman, G. W., Herbrandson, H. H., Indyke, B. H., Lohe, J. R., Ross, S., Sharma, D. D. and Shuster, W. W. (1982). *Modelling the fate of toxic organic materials in aquatic environments*. EPA-600/3-82-028, Athens, GA, 163 pp.

Payne, J. R. and Phillips, C. R. (1985). Photochemistry of petroleum in water, *Environ. Sci. Technol.* **19,** 569–79.

Perdue, E. M. and Lytle, C. R. (1983). Distribution model for binding of protons and metal ions by humic substances, *Environ. Sci. Technol.* **17,** 655–60.

Perdue, E. M. and Wolfe, N. L. (1983). Prediction of buffer catalysis in field and laboratory studies of pollutant hydrolysis reactions, *Environ. Sci. Technol.* **17,** 635–42.

Pereira, W. E., Rostad, C. E., Chiou, C. T., Brinton, T. I., Berber, L. B., II, Demcheck, D. K. and Demas, C. R. (1988). Contamination of estuarine water, biota, and sediment by halogenated organic compounds: A field study, *Environ. Sci. Technol.* **22,** 772–8.

Peterson, W. B. (1978). *User's Guide for PAL—A Gaussian Plume Algorithm for Point, Area, and Line Sources*. US EPA, Research Triangle Park, NC, Environ. Monit. Ser. EPA-600/4-78-013, NTIS PB 281–306.

Peyton, G. R. and Glaze, W. H. (1988). Destruction of pollutants in water with ozone in combination with ultraviolet radiation. 3. Photolysis of aqueous ozone, *Environ. Sci. Technol.* **22,** 761–7.

Pimentel, G. G. (1986). Ozone in the stratosphere, *Environ. Sci. Technol.* **20,** 328–9. (The third in a seven-part series.)

Pinder, G. F. (1984). Groundwater contaminant transport modelling, *Environ. Sci. Technol.* **18,** 108A–14A.

Pitzer, K. S. (1977). Electrolyte theory—Improvements since Debye and Hückel, *Acc. Chem. Res.* **10,** 371–7.

Pitzer, K. S. (1979). Theory: Ion interaction approach. In: *Activity Coefficients in Electrolyte Solutions,* Volume I, R. M. Pytkowicz (Ed.), CRC Press, Boca Raton, FL, 157–208.

Plankey, B. J. and Patterson, H. H. (1987). Kinetics of aluminium–fulvic acid complexation in acidic waters, *Environ. Sci. Technol.* **21,** 596–601.

Plankey, B. J. and Patterson, H. H. (1988). Effect of fulvic acid on the kinetics of aluminium fluoride complexation in acidic waters, *Environ. Sci. Technol.* **22,** 1454–9.

Platt, J. R. (1961). In: *Encyclopedia of Physics,* **37,** 173, S. Flugge (Ed.), Springer-Verlag, Berlin.

Podoll, R. T., Jaber, H. M. and Mill, T. (1986). Tetrachlorodibenzodioxin: Rates of volatilization and photolysis in the environment, *Environ. Sci. Technol.* **20,** 490–2.

Poisson, A. and Papaud, A. (1983). Diffusion coefficients of major ions in seawater, *Mar. Chem.* **13,** 265–80.

Pollard, J. E. and Hern, S. C. (1985). A field test of the EXAMS model in the Monongaheelea river, *Environ. Toxicol. Chem.* **4,** 361–9.

Popper, K. R. (1959). *The Logic of Scientific Discovery*. Hutchinson, London.

Post, J. (1978). Chemistry and the law: A review of recent developments, *Harvard Env. Law Rev.* **2,** 523–42.

Prahm, L. P. and Christensen, O. J. (1976). *Long-range transmission of sulphur pollutants computed by the pseudospectral model.* Danish Meteorological Institute, Air Pollution Section, Lyngbyvej, DK 2100. Prepared for the ECE Task Force for the Preparation of a Co-operative Programme for the Monitoring and Evaluation of the Long-range Transmission of Air Pollutants in Europe, October 1976, Lillestrom, Norway.

Prahm, L. P. and Christensen, O. J. (1977). Long-range transmission of pollutants simulated by a 2-dimensional pseudo-spectral dispersion model, *J. Appl. Meteorol.* **16,** 896–910.

Prausnitz, J. M. (1973). *The Molecular Thermodynamics of Fluid-Phase Equilibrium.* Prentice-Hall, Englewood Cliffs, NJ.

Prickett, T. A., Naymik, T. G. and Lonnquist, C. G. (1981). *A 'Random Walk' Solute Transport Model for Selected Groundwater Quality Evaluations.* Illinois State Water Survey, Bull. 65.

Pruppacher, H. R., Semonin, R. G. and Slinn, W. G. N. (1983). *Precipitation Scavenging, Dry Deposition, and Resuspension. Proc. 4th Int. Conf.,* Santa Monica, CA, Elsevier, New York.

Ramdahl, T. (1983). Polycyclic aromatic ketones in environmental samples, *Environ. Sci. Technol.* **17,** 666–70.

Randic, M. (1975). On characterization of molecular branching, *J. Am. Chem. Soc.* **97,** 6609–15.

Rao, P. S. C. and Jessup, R. E. (1982). Development and verification of simulation models for describing pesticide dynamics in soils, *Ecol. Modelling* **16,** 67–75.

Rathbun, R. E. and Tai, D. Y. (1986). Gas-film coefficients for the

volatilization of ethylene dibromide from water, *Environ. Sci. Technol.* **20,** 949–52.
Rawls, R. L. (1983). Dioxin's human toxicity is most difficult problem, *Chem. Eng. News.* **61,** 37–48.
Reid, G. K. and Wood, R. D. (1976). *Ecology of Inland Waters and Estuaries.* 2nd edn, D. Van Nostrand, New York.
Reid, R. C., Prausnitz, J. M. and Sherwood, T. K. (1977). *The Properties of Gases and Liquids.* 3rd edn, McGraw-Hill, New York, 544–601.
Reilly, P. M. (1970). Statistical methods in model discrimination, *Can. J. Chem. Eng.* **48,** 168–73.
Reimers, R. S. and Anderson, A. C. (1983). Environmental toxicology—physical chemical parameters, *J. Environ. Health* **45,** 288–90.
Reinhard, M., Barker, J. F. and Goodman, N. L. (1984). Occurrence and distribution of organic chemicals in 2 landfill leachate plumes, *Environ. Sci. Technol.* **18,** 953–61.
Rice, C. P., Samson, P. J. and Noguchi, G. (1986). Atmospheric transport of toxaphene to Lake Michigan, *Environ. Sci. Technol.* **20,** 1109–16.
Rich, V. (1986). Rhine pollution—Death of Europe sewer (Editorial), *Nature,* **324**(6094), 201.
Richardson, B. J. and Wild, J. S. (1982). Polychlorinated biphenyls (PCBs): An Australian viewpoint on a global problem, *Search,* **13**(1–2), 17–25.
Riggs, D. S. (1963). *The Mathematic Approach to Physiological Problems.* MIT Press, Cambridge, MA, p. 21.
Roberts, J. R. and Marshall, W. K. (1980). Retentive capacity: an index of chemical persistence expressed in terms of chemical-specific and ecosystem-specific parameters, *Ecotoxicol. Environ. Safety* **4,** 158–71. Also Appendix I: 281. NRCC 18157.
Rodhe, H. (1972). A study of the sulfur budget for the atmosphere over Northern Europe, *Tellus,* **24,** 128–38.
Rodhe, H. (1989). Acidification in a global perspective, *Ambio* **18,** 155–60.
Ross, R. D. and Crosby, D. G. (1975). The photo-oxidation of aldrin in water, *Chemosphere* **4,** 277–82.
Rowland, F. S. (1988). Chlorofluorocarbons, stratospheric ozone, and the Antarctic 'ozone hole', *Environ. Conserv.* **15,** 101–15.
Rowland, F. S. (1989). Chlorofluorocarbons and the depletion of stratospheric ozone, *Amer. Sci.* **77,** 36–45.
Ruedenberg, K. (1954). Free-electron network model for conjugated systems. V: Energies and electron distribution in the FE MO Model and in the LCAO MO Model, *J. Chem. Phys.* **22,** 1878–94.
Ruzic, I. (1987). Comment on 'Metal–humate interactions 1 and 2', *Environ. Sci. Technol.* **21,** 1132–5.
Rydholm, S. A. (1965). *Pulping Processes.* Interscience Publishers, New York.
Saar, R. A. and Weber, J. H. (1982). Fulvic acid: modifier of metal-ion chemistry, *Environ. Sci. Technol.* **16,** 510A–17A.
Saarikoski, J. and Viluksela, M. (1982). Relation between physicochemical properties of phenols and their toxicity and accumulation in fish, *Ecotoxicol. Environ. Safety* **6,** 501–12.

Sabljic, A. (1987). On the prediction of soil sorption coefficients of dynamic pollutants from molecular structure: Application of molecular topology model, *Environ. Sci. Technol.* **21,** 358–66.

Saleh, F. Y., Staples, C. A., Dickson, K. L. and Rodgers, J. H., Jr (1982). *Final report determining the compartmentalization and fate of chemicals in aquatic microcosms.* Chemical Manufacturers Assoc. Project ENV-14-W, May 1982, North Texas State Univ. Publication.

Sameshima, K., Simson, B. and Dence, C. W. (1979). Fractionation and characterization of toxic materials in Kraft spent bleaching liquors, *Svens. Papperstida.* **82,** 162–70.

Samiullah, Y. (1985). Biological effects of marine oil pollution, *Oil Petrochem. Pollut.* **2,** 235–64.

Sato, A. and Nakajima, T. (1979). A structure–activity relationship of some chlorinated hydrocarbons, *Arch. Environ. Health,* **34**(2), 69–75.

Sawyer, C. N. and Ryckman, D. W. (1957). Anionic synthetic detergents and water supply problems, *J. Am. Water Works Assoc.* **49,** 480.

Sax, N. I. and Lewis, R. J., Sr (1988). *Dangerous Properties of Industrial Materials.* 7th edn, 3 volumes, approx. 4000 pp.

Scatchard, G. (1949). *Ann. N.Y. Acad. Sci.* **51,** 660–72.

Scheff, P. A., Wadden, R. A. and Allen, R. J. (1984). Development and validation of a chemical element mass balance for Chicago, *Environ. Sci. Technol.* **18,** 923–31.

Scheibel, E. G. (1954). Liquid diffusivities, *Ind. Eng. Chem.* **46,** 2007–8.

Scheidigger, A. E. (1974). *The Physics of Flow Through Porous Media.* 3rd edn, University of Toronto Press, Toronto.

Schellenberg, K., Leuenberger, C. and Schwarzenbach, R. P. (1984). Sorption of chlorinated phenols by natural sediments and aquifer materials, *Environ. Sci. Technol.* **18,** 652–7.

Schiff, H. (1875). *Chem. Ber.* **8,** 1542.

Schnoor, I. L., Rao, N., Cartwright, K. J., Noll, R. M. and Ruiz-Calzada, C. E. (1983). *Verification of a toxic organic substance transport and bioaccumulation model.* EPA-600/3-83-007, Athens, GA, 164 pp.

Schnoor, J. L. (1981). Fate and transport of dieldrin in Coralville Reservoir: Residues in fish and water following a pesticide ban, *Science,* **211,** 840–2.

Schnoor, J. L. and McAvoy, D. C. (1981). Pesticide transport and bioconcentration model, *J. Environ. Eng. Div.* **107,** 1229–46.

Schrier, E., Pottle, M. and Scheraga, H. (1964). The influence of hydrogen and hydrophobic bonds on the stability of the carboxylic acid dimers in aqueous solutions, *J. Am. Chem. Soc.* **86,** 3444–9.

Schultz, D. (1982). Land disposal of hazardous waste. *Proc. 8th Ann. Res. Symp.,* J. F. Hutchell, Kentucky, 8–10 March 1982. Municipal Environ. Res. Lab. US EPA, Cincinnati, OH.

Schüürmann, G. and Klein, W. (1988). Advances in bioconcentration prediction, *Chemosphere* **17,** 1551–74.

Schwartz, F. W. and Growe, A. (1980). *A deterministic probabilistic model for contaminant transport.* US:NRC, NUREG/CR-1609, Washington, DC.

Schwille, F. (1981). In: *Quality of Groundwater,* W. van Diujvenbooden *et al.* (Eds), Elsevier, Amsterdam, 451–63.

Scow, K. M. (1982). Rate of biodegradation. In: *Handbook of Chemical Property Estimation Methods—Environmental Behaviour of Organic Compounds,* W. J. Lyman, W. F. Reehl and D. A. Rosenblatt (Eds), McGraw-Hill, New York (Chapter 9), 85 pp.

Seiber, J. N., Kim, Y. H., Wehner, T. and Woodrow, J. E. (1983). Analysis of xenobiotics in air. In: *Pesticide Residues and Formulation Chemistry, Proc. 5th Int. Congr. Pesticide Chem.,* Kyoto, Japan, 29 Aug.–4 Sept. 1982, J. Miyamoto and P. C. Kearney (Eds), Pergamon Press, Vol. 4, 3–12.

Shackelford, W. M. and Cline, D. M. (1986). Organic compounds in water, *Environ. Sci. Technol.* **20,** 652–7.

Shaeffer, D. L. (1980). A model evaluation methodology applicable to environmental assessment models, *Ecol. Modelling* **8,** 275–95.

Shah, J. J. and Singh, H. B. (1988). Distribution of volatile organic chemicals in outdoor and indoor air, *Environ. Sci. Technol.* **22,** 1381–8.

Shah, J. J., Johnson, R. L., Heyerdah, E. K. and Huntzick, J. J. (1986). Carbonaceous aerosol at urban and rural sites in the United States, *J. Air Pollut. Control Assoc.* **36,** 254–7.

Shaw, D. G. (1977). Hydrocarbons in the water column. In: *Fate and Effects of Petroleum Hydrocarbons in Marine Ecosystems and Organisms. Proc. Symp.,* 10–12 Nov. 1976, Seattle, WA. A. Wolfe (Ed.), Pergamon Press (Chapter 2), 8–18.

Shearer, R. C., Letey, J., Farmer, W. J. and Klute, A. (1973). Lindane diffusion in soil, *Soil Sci. Soc. Am. Proc.* **37,** 189–93.

Sheets, T. J. (1958). The comparative toxicities of four phenyl urea herbicides in several soil types, *Weeds* **6,** 413.

Sheih, C. H. (1977). Application of a statistical trajectory model to the simulation of sulfur pollution over north-east United States, *Atmos. Environ.* **11,** 173–8.

Shen, H. W. (Ed.) (1979). *Modelling of Rivers.* Wiley, New York.

Siegel, S. (1956). *Nonparametric Statistics for the Behaviour Sciences.* McGraw-Hill, New York, 312 pp.

Silka, L. R. and Swearingen, T. L. (1978). *A Manual for Evaluating Contamination Potential of Surface Impoundments.* US EPA, Ground Water Protection Branch, EPA570/9-78-003, 73 pp.

Simoneit, B. R. T. and Mazurek, M. A. (1981). Air pollution—the organic components, *CRC Crit. Rev. Environ. Control.* **11,** 219–76.

Simonin, H. A. (1988). Neutralization of acidic waters, *Environ. Sci. Technol.* **22,** 1143–5.

Singh, H. B. (1977*a*). Atmospheric halocarbons: Evidence in favour of reduced average hydroxyl radical concentrations in the troposphere, *Geophys. Res. Lett.* **4,** 101–4.

Singh, H. B. (1977*b*). Preliminary estimation of average tropospheric HO concentrations in the Northern and Southern Hemispheres, *Geophys. Res. Lett.* **4,** 453–6.

Singh, H. B. and Salas, L. J. (1979). Atmospheric halocarbons, hydrocarbons and sulfur hexafluoride: Global distribution, sources and sinks, *Science,* **203,** 899–903.

Singh, H. B., Salas, L. J., Shigeishi, H. and Smith, A. H. (1978). *Fate of halogenated compounds in the atmosphere—interim report.* EPA-600/3-78-017, January 1978.

Sjöström, E. (1981). *Wood Chemistry, Fundamentals and Applications.* Academic Press, New York.

Skiba, U., Cresser, M. S., Derwent, R. G. and Futty, D. W. (1989). Peat acidification in Scotland, *Nature* **337,** 68–9.

Slinn, W. G. N. (1978). *Relationship between removal processes and residence times for atmospheric pollutants.* NTIS CONF-780611-3, March 1978.

Smies, M. (1983). On the relevance of microecosystems for risk assessment: Some considerations for environmental toxicology, *Ecotoxicol. Environ. Safety* **7,** 355–65.

Smith, H. and White, T. (1929). The distribution ratios of some organic acids between water and organic liquids, *J. Phys. Chem.* **33,** 1953–74.

Smith, H. W. (1921). *J. Phys. Chem.* **25,** 204, 605.

Smith, J. H. *et al.* (1977/8). *Environmental pathways of selected chemicals in freshwater systems.* Part I: Background and experimental procedures—EPA-600/7/-77-113, 80 pp, Part II—EPA-600/7/-78/074, Athens, GA (Contract to SRI Int.).

Smith, J. H. and Bomberger, D. C. (1979). Prediction of volatilization rates of chemicals in water, *Water* 1978, *AIChE Symp. Ser. 190* **75,** 375–81.

Smith, J. H., Bomberger, D. C. and Haynes, D. L. (1980). Prediction of the volatilization rates of high volatility chemicals from natural water bodies, *Environ. Sci. Technol.* **14,** 1332–7.

Smolenski, E. A. (1964). Application of the theory of gases to calculations of the additive structural properties of hydrocarbons, *Russ. J. Phys. Chem.* **38,** 700–2.

Soil Conservation Service (1976). *Universal Soil Loss Equation.* Unnumbered technical note cited in Wagnet, Grenney, Woolridge and Jurinak, 1979.

Somayajulu, G. R. and Zwolinski, B. J. (1966). Generalized treatment of alkanes, *Trans. Faraday Soc.* **62,** 2327–40.

Somayajulu, G. R. and Zwolinski, B. J. (1972). Generalized treatment of alkanes, *J. Chem. Soc., Faraday Trans. II* **68,** 1971–87.

Spain, J. C., Pritchard, P. H. and Bourquin, A. W. (1980). Effects of adaptation on biodegradation rates in sediment/water cores from estuarine and freshwater environments, *Appl. Environ. Microbiol.* **40,** 726–34.

Spencer, W. F. (1970). Distribution of pesticides between soil, water and air. In: *Pesticides in the Soil: Ecology, Degradation and Movement,* Michigan State University, East Lansing, MI, 120–8.

Spencer, W. F. and Cliath, M. M. (1970). Desorption of lindane from soil as related to vapour density, *Soil Sci. Soc. Am. Proc.* **34,** 574–8.

Spencer, W. F., Farmer, W. J. and Cliath, M. M. (1973). Pesticide volatilization, *Residue Rev.* **49,** 1–47.

Sposito, G. (1981*a*). Trace metals in contaminated waters, *Environ. Sci. Technol.* **15,** 396–403.

Sposito, G. (1981*b*). *The Thermodynamics of Soil Solutions.* Clarendon Press, Oxford, UK.

Sposito, G. (1984). The future of an illusion: Ion activities in soil solutions, *Soil Sci. Soc. Amer. J.* **48,** 531–6.

Stainer, R. Y. (1948). The oxidation of aromatic compounds by fluorescent pseudomonads, *J. Bacteriol.* **55,** 477–94.

Strawinski, R. J. (1943). The dissimilation of pure hydrocarbons by members of the genus Pseudomonas. PhD Thesis, Pennsylvania State College, University Park, PA.

Stern, A. M. and Walker, C. R. (1978). Hazard assessment of toxic substances: environmental fate testing of organic chemicals and ecological effects testing. In: *Estimating the Hazard of Chemical Substances to Aquatic Life,* J. Cairns Jr, K. L. Dickson and A. W. Maki (Eds), Amer. Soc. Testing and Materials STP 657, Philadelphia, 81–131.

Stiver, W. and Mackay, D. (1984). Evaporation rate of spills of hydrocarbons and petroleum mixtures, *Environ. Sci. Technol.* **18,** 834–40.

Stolarski, R. S. (1988). The antarctic ozone hole, *Sci. Amer.* **258,** 20–6.

Stouch, R. T. and Jurs, P. C. (1985). Computer-assisted studies of molecular structure and genotoxic activity by pattern-recognition techniques, *Environ. Health Perpect.* **61,** 329–43.

Strand, R. H., Farrell, M. P., Goyert, J. C. and Daniels, K. L. (1983). Environment assessments through research data management, *J. Environ. Manage.* **16,** 269–80.

Stuper, A. J., Brugger, W. E. and Jurs, P. C. (1977). A computer system for structure–activity studies using chemical structure information handling and pattern recognition techniques, *Amer. Chem. Soc. Symp. Ser.* **52,** 165–91.

Stuper, A. J., Brugger, W. E. and Jurs, P. C. (1979). *Computer Assisted Studies of Chemical Structure and Biological Function.* Wiley-Interscience, New York.

Sugden, T. M. (1979*a*). The classification of pollutants and their pathways in the atmosphere, *Phil. Trans. R. Soc. Lond.* (*A*) **290,** 469–76.

Sugden, T. M. (Ed.) (1979*b*). *Pathways of pollutants in the atmosphere.* A Royal Society discussion organized by T. M. Sugden F.R.S., for the Royal Society's Study Group on Pollution in the Atmosphere, 3–4 Nov. 1977, The Royal Society, London, 166 pp.

Summers, P. W. (1983). A global perspective on acid deposition, its sources and atmospheric transport, *Water Qual. Bull.* **8,** 81–8, 109.

Swann, R. L. and Eschenroeder, A. (Eds) (1983). *Fate of chemicals in the environment: compartmental and multimedia models for predictions.* Symp. on fate of chemicals in the environment at the 184th Meeting of the Amer. Chem. Soc., Kansas City, MO, ACS. Symp. Ser. 225.

Swann, R. L., McCall, P. J. and Unger, S. M. (1979). *Volatility of pesticides from soil surfaces.* Unpublished undated manuscript, received by Thomas, R. G. (1982) as a personal communication from Dow Chemical USA, Midland, MI, 16 November 1979.

Swann, R. L., McCall, P. J., Laskowski, D. A. and Dishburger, H. J. (1981). Estimation of soil sorption constants of organic chemicals by high-performance liquid chromatography. In: *Aquatic Toxicology and Hazard Assessment*; *4th Conf.,* D. R. Branson and K. L. Dickson (Eds), Am. Soc. for Testing and Materials, 1981, ASTM STP 737, 43–8.

Swindoll, C. M., Aelion, C. M. and Pfaender, F. K. (1988). Influence of mineral and organic nutrients on aerobic biodegradation and the adaptation response of subsurface microbial communities, *Appl. Environ. Microbiol.* **54,** 212–17.

Swisher, R. D. (1963*a*). Biodegradation of alkylbenzene sulfonates in relation to chemical structure, *J. Water Pollut. Control Fed.* **35,** 877–92.

Swisher, R. D. (1983*b*). Biodegradation rates of isomeric diheptylbenzene sulfonates, *Dev. Ind. Microbiol.* **4,** 39–45.

Swisher, R. D. (1970). *Surfactant Biodegradation.* Marcel Dekker, New York.

Sylvester, J. J. (1874). On an application of the new atomic theory to the graphical representation of the invariants and covariants of binary quantics—with three appendices, *Am. J. Math.* **1,** 64–128.

Tabak, H. H., Chambers, C. W. and Kabler, P. W. (1964). Microbial metabolism of aromatic compounds: I. Decomposition of phenolic compounds and aromatic hydrocarbons by phenol-adapted bacteria, *J. Bacteriol.* **87,** 910.

Tang, D. H., Schwartz, F. W. and Smith, L. (1982). Stochastic modeling of mass transport in a random velocity field, *Water Resour. Res.* **18,** 231–44.

Tanii, H. and Hashimoto, K. (1982). Structure–toxicity relationship of acrylates and methacrylates, *Toxicol. Lett.* **11,** 125–9.

Thomann, R. V., Szumski, D. S., Ditoro, D. M. and O'Connor, D. J. (1974). A food chain model of cadmium in Western Lake Erie, *Water Res.* **8,** 841–9.

Thomas, R. G. (1982*a*). Volatilization from soil. In: *Handbook of Chemical Property Estimation Methods—Environmental Behaviour of Organic Compounds,* W. J. Lyman, W. F. Reehl and D. H. Rosenblatt (Eds), McGraw-Hill (Chapter 16), 50 pp.

Thomas, R. G. (1982*b*). Volatilization from water. In: *Handbook of Chemical Property Estimation Methods—Environmental Behaviour of Organic Compounds,* W. J. Lyman, W. F. Reehl and D. H. Rosenblatt (Eds), McGraw-Hill (Chapter 15), 34 pp.

Tiwari, J. L. and Hobbie, J. E. (1976). Random differential equations as models of ecosystems—II. Initial conditions and parameters specifications in terms of maximum entropy distributions, *Math. Biosci.* **31,** 37–53.

Tomovic, R. (1963). *Sensitivity Analysis of Dynamic Systems.* McGraw-Hill, New York, 142 pp.

Tsivoglou, E. C. and Neal, L. A. (1976). Tracer measurement of re-aeration. III: Predicting the re-aeration capacity of inland streams, *J. Water Pollut. Control Fed.* **48,** 2669–99.

Tsivoglou, E. C., O'Connell, R. L., Walter, C. M., Godsil, P. J. and Logsdon, G. S. (1965). Tracer measurements of atmospheric re-aeration. I: Laboratory studies, *J. Water Pollut. Control Fed.* **37,** 1343–62.

Tsivoglou, E. C., Cohen, J. B., Shearer, S. D. and Godsil, P. J. (1968). Tracer measurement of stream re-aeration. II: Field studies, *J. Water Pollut. Control Fed.* **40,** 285–305.

Tucker, S. P. and Carson, G. A. (1985). Deactivation of hazardous chemical wastes, *Environ. Sci. Technol.* **19,** 215–20.

Tucker, W. A. and Nelken, L. H. (1982). Diffusion coefficients in air and water. In: *Handbook of Chemical Property Estimation Methods—Environmental Behaviour of Organic Compounds,* W. J. Lyman, W. F. Reehl and D. H. Rosenblatt (Eds), McGraw-Hill (Chapter 17), 25 pp.

Turner, D. B. (1979). Atmospheric dispersion modelling. A critical review, *Air Pollut. Control Assoc. J.* **29,** 502–19.

Turner, D. B. and Busse, A. D. (1973). *User's guide to the interactive versions of three point source dispersion programs: PTMAX, PTDIS, and PTMTP.* Preliminary Draft, Meteorology Lab., US EPA, Research Triangle Park, NC, 1973. (Cited in Turner (1979).)

Turner, D. B. and Novak, J. H. (1978). *User's Guide for RAM.* EPA-600/8-78-016, U.S. EPA, Research Triangle Park, NC, November 1978.

UKM/USM/SRC (1986). *Biodegradation of crude oil in Sabah and Sarawak marine environment.* (A study conducted for Petroleum National Berhad (PETRONAS), Sabah Shell Petroleum Company and Sarawak Shell Berhad), University Kebangsaan Malaysia, Universiti Sains Malaysia and Shell Sittingbourne Research Centre, UK, 221 pp.

UNSCEAR (1982). *Ionizing radiation: sources and biological effects.* 1982 Report to the General Assembly, United Nations Scientific Committee on the Effects of Atomic Radiation, with annexes, UN publication E.82.IX.8, New York.

UNSCEAR (1989). *Exposures from the Chernobyl accident.* United Nations Scientific Committee on the Effects of Atomic Radiation, 214 pp and annexes.

Valvani, S. C. and Yalkowsky, S. H. (1980). In: *Physical Chemical Properties of Drugs,* S. H. Yalkowsky, A. A. Sinkula and S. C. Valvani (Eds), Marcel Dekker, New York.

van Duyne, R., Taylor, S. Christian, S. and Affsprung, H. (1967). Self-association and hydration of benzoic acid in benzene, *J. Phys. Chem.* **71,** 3427–30.

van Noort, P. C. M. and Wondergem, E. (1985). Scavenging of airborne polycyclic aromatic hydrocarbons by rain, *Environ. Sci. Technol.* **19,** 1044–8.

Varanasi, U., Reichert, W. L., Stein, J. E., Brown, D. W. and Sanborn, H. R. (1985). Bioavailability and biotransformation of aromatic hydrocarbons in benthic organisms exposed to sediment from an urban estuary, *Environ. Sci. Technol.* **19,** 836–41.

Veith, G. D. and Kosian, P. (1983). Estimating bioconcentration potential from octanol/water partition coefficients. In: *Physical Behaviour of PCBs in the Great Lakes,* D. Mackay, S. Paterson and St. J. Eiseureich (Eds), Ann Arbor Science Publications, Ann Arbor, MI, 269–82.

Veith, G. D., Austin, N. M. and Morris, R. T. (1979*a*). Rapid method for estimating log *P* for organic chemicals, *Water Res.* **13,** 43–7.

Veith, G. D., Defoe, D. L. and Bergstedt, B. V. (1979*b*). Measuring and estimating the bioconcentration factor of chemicals in fish, *J. Fish Res. Board Can.* **36,** 1040–8.

Veith, G. D., Macek, K. J., Petrocelli, S. R. and Carroll, J. (1980*a*). An evaluation of using partition coefficients and water solubilities to estimate bioconcentration factors for organic chemicals in fish. In: *Aquatic Toxicology,* J. R. Eaton, P. R. Parrish and A. C. Hendricks (Eds), ASTS, Philadelphia, 116–29.

Veith, G. D., Macek, K. J., Petrocelli, S. R. and Carroll, J. (1980*b*). Preprint,

J. Fish. Res. Board Canada, 1980. (Cited after W. J. Lyman, W. F. Reehl and D. H. Rosenblatt, 1982 *Handbook of Chemical Properties Estimation Methods,* McGraw-Hill Book Co., pp. 5-4, 5-30.)

Venezian, E. C. (1976). *Pre-screening for environmental hazards—A system for selecting and prioritizing chemicals.* Report to the US EPA by Arthur D. Little Inc. (June 1976).

Venkatram, A. (1980). Dispersion from an elevated source in a convective boundary layer, *Atmos. Environ.* **14,** 1–10.

Venkatram, A. (1981). Model predictability with reference to concentrations associated with point sources, *Atmos. Environ.* **15,** 1517–22.

Venkatram, A. and Karamachandani, P. (1986). Source–receptor relationships, *Environ. Sci. Technol.* **20,** 1084–91.

Venkatram, A. and Vet, R. (1981). Modelling of dispersion from tall stacks, *Atmos. Environ.* **15,** 1531–8.

Ventullo, R. M. and Larson, R. J. (1986). Adaptation of aquatic microbial communities to quaternary ammonium compounds, *Appl. Environ. Microbial.* **51,** 356–61.

Vogel, T. M., Criddle, C. S. and McCarthy, P. L. (1987). Transformations of halogenated aliphatic compounds, *Environ. Sci. Technol.* **21,** 722–36.

Voorhees, M. L. (1988). *Intersat.* US Geological Survey (computer model).

Voorhees, M. L. and Rice, J. L. (1988). *Intertrans.* US Geological Survey (computer model).

Voss, C. I. (1988). *SUTRA* (*Saturated–Unsaturated Transport*). US Geological Survey and the Engineering and Services Laboratory, US Air Force Engineering and Services Center.

Voss, R. H., Wearing, J. T. and Wong, A. (1981). *Advances in the Identification and Analysis of Organic Pollutants in Water,* L. H. Keith (Ed.), Ann Arbor Science Publications, Ann Arbor, MI.

Wada, R. Y., Leong, E. Y. and Robinson, L. H. (1979). A method for analysing alternative oxidant control strategies, *J. Air Pollut. Control Assoc.* **29,** 346–51.

Wagnet, R. J., Grenney, W. J. and Jurinak, J. J. (1978). Environmental transport model of heavy metals, *J. Environ. Eng. Div. ASCE,* **104,** 61–77.

Wagnet, R. J., Grenney, W. J., Woolridge, G. L. and Jurinak, J. J. (1979). An atmospheric–terrestrial heavy metal transport model. I. Model Theory, *Ecol. Modelling* **6,** 253–72. (II. Process equations in Grenney *et al.* (1979).)

Waite, T. D., Wrigley, I. C. and Szymczak, R. (1988). Photoassisted dissolution of a colloidal manganese oxide in the presence of fulvic acid, *Environ. Sci. Technol.* **22,** 778–85.

Walden, C. C. (1976). Toxicity of pulp and paper-mill effluents and corresponding measurement procedures, *Water Res.* **10,** 639–64.

Walden, C. C. and Howard, T. E. (1977). Toxicity of pulp and paper-mill effluents—Review of regulations and research, *Tappi* **60,** 122–5.

Walker, A. and Crawford, D. V. (1970). Diffusion coefficients for two triazine herbicides in six soils, *Weed Res.* **10,** 126–32.

Weathers, K. C., Likens, G. E., Bormann, F. H., Bicknell, S. H., Bormann, B. T., Daube, B. C., Jr, Eaton, J. S., Galloway, J. N., Keene, W. C.,

Kimball, K. D., McDowell, W. H., Siccama, T. G., Smiley, D. and Tarrant, R. A. (1988). Cloudwater chemistry from ten sites in North America, *Environ. Sci. Technol.* **22,** 1018–26.

Webley, D. M., Duff, R. B. and Farmer, V. C. (1959). Effect of substitution in the side-chain on beta-oxidation of aryloxyalkylcarboxylic acid by nocardia-opaca, *Nature* (*London*) **183,** 748–9.

Weimer, R. F. and Prausnitz, J. M. (1965). *Hydrocarbon Process.* **44,** 237–42.

Westcott, J. W., Simon, C. G. and Bidleman, T. F. (1981). *Environ. Sci. Technol.* **11,** 1375–8.

Whelpdale, D. M. (1982). Deposition processes. In: *United States–Canada Memorandum of Intent on Transboundary Air Pollution,* Atmospheric Sciences and Analysis Work Group 2. Report 2F. Atmospheric Environment Service, Toronto, Canada.

Whelpdale, D. M. (1983). Acid deposition: distribution and impact, *Water Qual. Bull.* **8,** 72–80, 109.

Whelpdale, D. M. and Munn, R. E. (1976). Global sources, sinks, and transport of air pollution. In: *Air Pollution.* 3rd edn, Vol. 1: *Air Pollutants, their Transformation and Transport,* Academic Press, New York (Chapter 7), 289–324.

Whitby, K. T. (1978). Physical characteristics of sulfur aerosols, *Atmos. Environ.* **12,** 135–59.

Whitfield, M. (1979). Activity coefficients in natural waters. In: *Activity Coefficients in Electrolyte Solutions,* Vol. II, R. M. Pytkowicz (Ed.), CRC Press, Boca Raton, FL, 153–299.

Whitman, W. G. (1923). *Chem. Metall. Eng.* **29,** 146.

WHO (1979). *Sulphur oxides and suspended particulate matter.* Environmental Health Criteria Document No. 8, World Health Organization, Geneva.

WHO (1987). *Technical Manual for the Safe Disposal of Hazardous Wastes with Special Emphasis on the Problems and Needs of Developing Countries.* World Health Organization, United Nations Environment Programme and the World Bank, prepared for restricted distribution by Environmental Resources Ltd, 4 volumes (PEP/87.2).

Wiener, H. (1947). Correlation of heats of isomerization, and differences in heats of vaporization of isomers, among the paraffin hydrocarbons, *J. Am. Chem. Soc.* **69,** 2636–8.

Wiener, H. (1948*a*). Relation of the physical properties of the isomeric alkanes to molecular structure, *J. Phys. Chem.* **52,** 1082–9.

Wiener, H. (1948*b*). Vapour pressure–temperature relationships among the branched paraffin hydrocarbons, *J. Phys. Chem.* **52,** 425–30.

Wiggins, B. A., Jones, S. H. and Alexander, M. (1987). Explanations for the acclimation period preceding the mineralization of organic chemicals in aquatic environments, *Appl. Environ. Microbiol.* **53,** 791–6.

Wilke, C. R. and Chang, P. (1955). Correlation of diffusion coefficients in dilute solutions, *AIChE J.* **1,** 264–70.

Williamson, B. and Craig, L. (1947). Identification of small amounts of organic compounds by distribution studies. V: Calculation of theoretical curves, *J. Biol. Chem.* **168,** 687–97.

Wischmeier, W. H. (1976). Use and misuse of the universal soil loss equation, *J. Soil Water Conserv.* (Feb.): 5–9.

Wise, H. E., Jr and Fahrenthold, P. D. (1981). Predicting priority pollutants from petrochemical processes, *Environ. Sci. Technol.* **15,** 1292–304.

Wold, S. (1976). Pattern recognition by means of disjoint principal components models, *Pattern Recog.* **8,** 127–39.

Wold, S. and Sjostrom, M. (1977). SIMCA: A method for analyzing chemical data in terms of similarity and analogy, *Amer. Chem. Soc. Symp. Ser.* **52,** 243–82.

Wolfe, D. A. (Ed. and others) (1977). Fate and effects of petroleum hydrocarbons in marine organisms and ecosystems. *Proc. Symp. November 1976, Seattle, Washington,* Pergamon Press, Oxford, 478 pp.

Wolff, P. M., Hansen, W. and Joseph, J. (1972). Investigation and predication of dispersion of pollutants in the sea with hydrodynamical numerical (HN) models. In: *Marine Pollution and Sea Life,* M. Ruivo (Ed.), FAO/Fishing News (Books) Ltd., 625 pp. From FAO Tech. Conf. on Mar. Pollut. and its effects on living resources and fishery, Rome, 9–18 Dec. 1970. 146–50.

Wood, W. P. (1981). *Comparison of environmental compartmentalization approaches.* OECD Chemicals Group, Working Party of Exposure Analysts (EXPO), Room Doc. 80.21.

Woodburn, K. B., Doucette, W. J. and Andrew, A. W. (1984). Generator column determination of octanol/water partition coefficients for selected polychlorinated biphenyl congeners, *Environ. Sci. Technol.* **18,** 457–9.

Woodrow, J. E., Seiber, J. N. and Kim, Y.-H. (1986). Measured and calculated evaporation losses of two petroleum hydrocarbon herbicide mixtures under laboratory and field conditions, *Environ. Sci. Technol.* **20,** 783–9.

Wroth, B. and Reid, E. (1916). *J. Am. Chem. Soc.* **38,** 2316.

Wu, Jy, S. and Hilger, H. (1984). Evaluation of EPA's Hazard Ranking System, *J. Environ. Eng.* **110,** 797–807.

Wulf, R. J. and Featherstone, R. M. (1957). A correlation of Van der Waals' constants with anesthetic potency, *Anesthesiology* **18,** 97–105.

Wyndham, R. C. (1986). Evolved aniline catabolism in *Acinetobacter calcoaccticus* during continuous culture in river water, *Appl. Environ. Microbiol.* **51,** 781–9.

Yalkowsky, S. H. (1979). Estimation of eutropies of fusion of organic compounds, *Ind. Eng. Chem. Fundam.* **18,** 108–11.

Yalkowsky, S. H. and Valvani, S. C. (1979). Solubilities and partitioning. II: Relationships between aqueous solubilities, partition coefficients and molecular surface areas of rigid aromatic-hydrocarbons, *J. Chem. Eng. Data,* **24,** 127–9.

Yalkowsky, S. H., Valvani, S. C. and Amidon, G. L. (1976). Solubility of non-electrolytes in polar solvents. IV: Non-polar drugs in mixed solvents, *J. Pharm. Sci.* **65,** 1488–94.

Yeh, G. T. and Ward, D. S. (1981). *FEMWASTE: A finite-element model of waste transport through saturated–unsaturated porous media.* Oak Ridge Natl. Lab., Environ. Sci. Div., Publ. No. 1462. ORNL-5601, 137 pp.

Yoshida, K., Shigeoka, T. and Yamauchi, F. (1983). Non-steady-state equilibrium model for the preliminary prediction of the fate of chemicals in the environment, *Ecotox. Environ. Safety,* **7,** 179–90.

Yosie, T. F. (1987). EPA's risk assessment culture, *Environ. Sci. Technol.* **21,** 526–31.

Yost, K. J., Miles, L. J. and Greenkorn, R. A. (1979). Cadmium in the environment: A systems analysis overview. In: *Trace Substances in Environmental Health XII.* Proc. Conf. Trace Substances in Environ. Health, Columbia, MS, June 1979, D. D. Hemphill (Ed.), p. 235.

Young, P. (1983). Systems methods in the evaluation of environmental pollution problems. In: *Pollution: Causes, Effects and Control,* R. M. Harrison (Ed.), RSC Spec. Publ. No. 44, 322 pp. (Chapter 15), 258–76.

Zafiriou, O. C. (1984). *Bibliography of natural water photochemistry.* Woods Hole Oceanographic Institution Technical Memorandum, 2–84.

Zafiriou, O. C., Joussot-Dubien, J., Zepp, R. G. and Zika, R. G. (1984). Photochemistry of natural waters, *Environ. Sci. Technol.* **18,** 358A–71A.

Zaroogian, G. E., Heltshe, J. F. and Johnson, M. (1985). Estimation of bioconcentration in marine species using structure activity models, *Environ. Toxicol. Chem.* **4,** 3–12.

Zepp, R. G. and Baughman, G. L. (1978). In: *Aquatic Pollutants: Transformation and Biological Effects,* O. Hutzinger, I. H. Van Lelyveld and B. C. J. Zoeteman (Eds), Pergamon Press, Oxford, 237–63.

Zepp, R. G. and Cline, D. M. (1977). Rates of direct photolysis in aquatic environment, *Environ. Sci. Technol.* **11,** 359–66.

Zepp, R. G. and Schlotzhauer, P. F. (1979). In: *Polynuclear Aromatic Hydrocarbons,* Third International Symposium on Chemistry and Biology—Carcinogenesis and Mutagenesis, P. W. Jones and P. Labor (Eds), Ann Arbor Science Publications, Ann Arbor, MI, 141–58.

Zepp, R. G., Baughman, G. L. and Schlotzhauer, P. F. (1980). *Photosensitization of pesticide reactions by humic substances,* 2nd Chem. Congr. of the North American Continent, San Francisco, August 1980.

Zepp, R. G., Wolfe, N. L., Azarraga, L. V., Cox, R. H. and Pape, C. W. (1977*a*). Photochemical transformation of the DDT and methoxychlor degradation products, DDE and DMDE by sunlight, *Arch. Environ. Contam. Toxicol.* **6,** 305–14.

Zepp, R. G., Wolfe, N. L., Baughman, G. L. and Hollis, R. C. (1977*b*). Singlet oxygen in natural waters, *Nature* **267,** 421–3.

Zepp, R. G., Schlotzhauer, P. F. and Sink, R. M. (1985). Photosensitized transformations involving electronic energy transfer in natural waters: Role of humic substances, *Environ. Sci. Technol.* **19,** 74–81.

Zepp, R. G., Braun, A. M., Hoigné, J. and Leenheer, J. A. (1987*a*). Photoproduction of hydrated electrons from natural organic solutes in aquatic environments, *Environ. Sci. Technol.* **21,** 485–90.

Zepp, R. G., Hoigné, J. and Bader, H. (1987*b*). Nitrate-induced photooxidation of trace organic chemicals in water, *Environ. Sci. Technol.* **21,** 443–50.

Zimmerman, J. R. and Thompson, R. S. (1975). *User's guide for HIWAY: A highway air pollution model.* US EPA, Research Triangle Park, NC, Environ. Monit. Ser., EPA-650/4-74-008, NTIS PB 239–944.

Zoeteman, B. C. J. (1973). *The potential pollution index as a tool for river water quality management*, August 1973, Technical Paper No. 6, Government Institute for Water Supply, World Health Organization, International Reference Centre for Community Water Supply, The Hague, The Netherlands.

Index